SOCIAL BEHAVIOUR IN ANIMALS

SOCIAL BEHAVIOUR IN ANIMALS

Dr. S.K. Singh

RANDOM PUBLICATIONS

NEW DELHI - 110 002 (INDIA)

Social Behaviour in Animals

ISBN 978-93-51116-89-9

Published in 2015 in India by

RANDOM PUBLICATIONS

4376-A/4B, Gali Murari Lal, Ansari Road
New Delhi-110 002
Phone: +9111-43580356, 23289044
E-mail: randomexports@gmail.com; sales@randompublications.com; info@randompublications.com

Reprinted 2025

Type Setting by: Friends Media, Delhi-110089
Digitally Printed at: Replika Press Pvt. Ltd.

Preface

Animal behaviour, the concept, broadly considered, referring to everything animals do, including movement and other activities and underlying mental processes. Human fascination with animal behaviour probably extends back millions of years, perhaps even to times before the ancestors of the species became human in the modern sense. Initially, animals were probably observed for practical reasons because early human survival depended on knowledge of animal behaviour. Whether hunting wild game, keeping domesticated animals, or escaping an attacking predator, success required intimate knowledge of an animal's habits. Even today, information about animal behaviour is of considerable importance. Evolution has selected animals for increased fitness in their natural environments over thousands of generations. Fitness in this context can roughly be defined as reproductive success, and is not necessarily related to welfare; an animal with high fitness may have a poorer state of welfare than one with low fitness. One of the major achievements of behavioural research over the last decades is the realisation that animals are equipped with flexible behavioural programmes, usually referred to as strategies. A strategy involves a number of different possible behavioural outputs and a decision mechanism that helps the animal in selecting the best behaviour under different conditions. The best strategy in this context is the strategy that gives the largest possible fitness for the animal.

During domestication, animals have been genetically changed in relation to their ancestors, which has affected various aspects of their adaptive traits, including behaviour. However, as is obvious from many comparisons between domestic animals and their ancestors, behaviour is usually affected in a very limited manner. Generally, no new behaviours have developed and none have completely disappeared from the gene pool. The changes are usually referred to as altered release thresholds, but a better description may be to say that the strategies have been modified. Since domestication provides animals with food and protection, and involves selection for specific traits, the optimal behaviour is likely to be different compared to when food is scarce and predation pressure high. Other strategies than in the wild may therefore give the highest fitness. We have found that domestic pigs and poultry use slightly different strategies than their wild counterparts, wild boars and jungle fowl, during food search, and the differences appear to reflect an adaptation of strategies to the conditions offered during domestication As behaviour becomes more complex in the animal kingdom, a given species develops several alternate patterns of behaviour for each general function. During domestication, animals have been genetically changed in relation to their ancestors, which have affected various aspects of their adaptive traits, including behaviour. Generally, no new behaviours have developed and none have completely disappeared from the gene pool. The changes are usually referred to as altered release thresholds, but a better description may be to say that the strategies have been modified. Since domestication provides animals with food and protection, and involves selection for specific traits, the optimal behaviour is likely to be different compared to when food is scarce and predation pressure high.

An understanding of social behaviour is increasingly necessary in farm animal husbandry as more animals are housed in groups rather than in individual stalls or pens. There may be economic or welfare reasons for such housing. This book is one of the few to specifically address this important subject.

I thank all members of my team who have helped in the preparation of the book. My special thanks go to "Random Publications" who have published the book.

— Dr. S.K. Singh

Contents

Chapter 1

Natural Selection in Animal Behaviour

More offspring are produced than finite resources can support. Individuals thus can be viewed as in a constant struggle for existence. Individuals within a population are rarely clones, they commonly show variation in phenotypes as well as genotypes. Some of this variation in behavioural phenotypes is heritable. As successful variants are more likely to survive and reproduce, their genotypes will be become overrepresented in the next generation

Density-independent and density-dependent growth models: Exponential Model: a species can potentially increase in numbers according to a geometric series — Thomas Robert Malthus (1766-1834) Logistic Model: the rate of population increase may be limited, i.e. it may depend on population density — Pierre Verhulst (1838). Carrying Capacity (K): an environment's maximum persistently supportable load (Catton 1986).

Natural selection is the process by which environmental effects lead to varying degrees of reproductive success among individuals of a population of organisms with different hereditary characters, or traits. The characters that inhibit reproductive success decrease in frequency from generation to generation. It is the process whereby certain genes (alleles) gain greater representation in the following generations compared to other alleles. Adaptations are the complement of traits that increases the fitness of the owner. An individual's Fitness or Reproductive Success is the relative probability that an animal of a particular genotype and phenotype will manage to contribute its genes to the next generation

Aside from Natural Selection, changes in gene frequencies within a population may also arise from a variety of other sources including:

- *Mutations:* changes to individual nucleotide bases along the DNA to largescale rearrangements of chromosomes
- *Gene flow:* Emigration, Immigration
- Genetic drift: Changes in the genetics of the population as a result of chance. Bottlenecks and Founder Effect: Ellis-van Creveld Syndrome is an autosomal recessive skeletal dysplasia that results in short-limbed disproportionate dwarfism. The syndrome most commonly occurs in the Amish population of Lancaster, Pa. The incidence is approximately 5 per 1000 live births and 2 per 1000 living persons. No case had been described in the Amish of Ohio, Indiana, or other Amish areas at the time of an extensive search performed by McKusick, et al. The genealogy of the disorder in the Amish of Lancaster, Pa. can be traced back to the immigrants Samuel King and his wife

Example: H.B.D. Kettlewell's work on Industrial Melanism

Sexual Selection and Mate Choice by Females

Sexual selection, a subcategory of natural selection, was first recognised by Charles Darwin and "occurs when individuals differ in their ability to compete with others for mates or to attract members of the opposite sex" (Alcock 493). By heavy courtship, fighting, or large territorial possession, males heavily compete for females. Eventhough a male may win a fierce competition for the mate of his choice, it is ultimately the female who decides on a partner that she wants. The female is often successful in her attempts to control reproduction by being choosy and having particular preferences for a male mate.

Females choose mates based on many factors. One important factor is male adornments, or ornaments. For example, Marion Petrie and Tim Halliday concluded that the removal of eyespots from a peacock's tail significantly reduces his attractiveness to females. "After 20 eyespots had been cut from their tails, males averaged two fewer mates in the following breeding season compared with their performance in the previous year" (Alcock 348). Thus, females are extremely conscious of the visual stimuli provided by males. Elaborate sensory cues alert the female that the male is reproductively superior to others. "Male adornments may more readily elicit the mating responses of some females who will thus mate preferentially in favour of the adorned males" (Bateson 53). Other factors involved in female preference in mate selection include body colouration or the "gift" that the male may present to the female before copulation.

Four theories are used to explain mate choice in females. The good parent theory suggests that "choosy individuals select partners on the basis of how well they will care for their offspring" (Alcock 491). This theory focuses on the female's search of a paternal male. The healthy mate theory occurs when females prefer "males healthy enough to produce and maintain elaborate ornaments" (Alcock 491). A good example of this is in female house finches, who choose male mates based on their bright colouration. Bright colouration tells the female that the male is more resistant to pathogens and parasites.

The good genes theory "argues that females exhibit mate choice in order to provide their offspring with a partner's genes that will advance their offspring's chances of survival or reproductive success" (Alcock 491). A mother always wants the best for her child, even if it is a future child. Females select mates with certain traits, because they want their children to be healthy, viable, and reproductively successful. "Females that mate with attractive males are compensated for reduced fecundity by bearing "sexy sons" with higher than average mating success" (Andersson 45). The least obvious theory is the runaway selection theory. It states that by being choosy, females "create a positive feedback loop favoring both males with these attributes and females that prefer them" (Alcock 493).

A direct benefit of mate choice by females is the assurance of bearing offspring that will survive well and display high general fitness. This is a heritable benefit. Nongenetic benefits of mate choice include fecundity advantages, food, parental care, or a good territory. All of the benefits, both heritable and nonheritable, ultimately lead to the greater survival of a female's offspring.

Cuckoo

Brood Parasitism

Common cuckoos (*Cuculus canorus*) in Europe, and brown-headed cowbirds (*Molothrus ater*) in North America lay their eggs in the nests of other species and leave them to the care of their respective host. A Cuckoo female searches for an unguarded nest of another species and then replaces an egg in it with one of her own. The mother then abandons her egg leaving it to be raised by the nest's natural owners. Clearly, lacking a distinct ability to identify and remove foreign eggs from among its clutch appears maladaptive.

As an example of an evolutionary arms race, cukoos have adapted their kleptoparasitic strategies to the host behaviour. Upon return, hosts may reject the foreign egg if it is a poor match, added at the wrong time of day, or the cukoo's presence has been witnessed. Different cuckoo strains exhibit distinct preferences for a specific host species which provide a close visual match to their own eggs.

Likewise, hosts have adapted to selection from cuckoos where species suffering high amounts of parasitism are much better at identifying and removing cuckoo eggs than those who suffer little. When parents remove an egg from their nest they do risk removing one of their own or breaking some of the others in the process. Whether taking such risks will be worth it depends on the overall rate of parasitism and the ability to discriminate foreign from own eggs.

In contrast to cuckoos, brownheaded cowbirds exhibit little preference for nests of a specific song bird species. Moreover, their eggs fail to show close mimicry to those of their hosts. Removal of parasitic eggs is much more variable, high in some species and rare in others. Although cowbird fledglings do not evict their nest mates, up to 50% of the nests may be parasitised in some areas.

The lack of co-evolution in brood parasitism of cowbirds compared to that of cuckoos is puzzling. Do these two cases represent different stages of the arms race? Are cowbirds in the early parts of the game while cuckoos have had a chance to adapt for a much longer time? Has the recent change in North America landscape from forest to agricultural lands brought cowbirds into contact with a much greater number of new hosts? Alternatively, both species may be at an evolutionary equilibrium with stabilising selection. Differing degrees of rejection may well depend on their respective costs of rejecting vs. accepting any given egg.

Operant Conditioning

Operant conditioning, sometimes called instrumental conditioning or instrumental learning, was first extensively studied by Edward L. Thorndike (1874-1949). Thorndike's most famous work investigated the behaviour of cats trying to escape from various home-made puzzle boxes. When first constrained in the boxes the cats took a long time to escape from each. With experience however, ineffective responses occurred less frequently and successful responses occurred more quickly enabling the cats to escape in less and less time over successive trials.

B.F. Skinner (1904-1990) extended the theories proposed by Thorndike about 40 years after Thorndike published his works.

Skinner's ideas of animals operating on the environment led him to analyse how behaviour is affected by its consequences. The contraption that became known as the Skinner box measured whether a task was completed and how long it took for the task to be completed. The Skinner box consisted of a bar that, when pressed, triggered the release of a food pellet.

Skinner believed that rewarding an animal when an appropriate action occurred would increase the likelihood that the behaviour would be repeated. When a rat accidentally tapped on the bar, a food pellet was released, and Skinner observed the amount of time it took the rat to find the pellet. As the rat learned that a pellet released each time he happened to step on the bar, the rat learned to press the bar and to immediately find the food. This training became known as operant conditioning (Alcock).

Today Skinners operant conditioning is still used to train animals. All animals that are part of Sea World programs have been trained using positive reinforcement operant conditioning. When an animal performs a desired behaviour the trainer will reward the animal to encourage the action to be performed when signaled for at a later time. Positive reinforcers are stimuli that strengthen a response if they follow that response (How Animals). Negative reinforcers can also be used to encourage a certain behaviour. It is important to distinguish the difference between negative reinforcement and the punishment theory. Negative reinforcement is simply the removal of an unpleasant stimulant or pain (Bernstein). For example: a person who uses an alarm that plays a pleasant song may oftentimes hit the snooze or linger in bed.

To condition himself to get out of bed as soon as the alarm goes off the person may switch the alarm sound to something obnoxiously loud and unpleasant. To remove the unpleasant stimulant the person will immediately get out of bed and turn the alarm off. Knowing how to remove the unpleasant sound increases the likelihood that the person will get out of bed immediately after the alarm plays. The Punishment theory, on the other hand, introduces an aversive stimulus to reduce the likelihood of a certain behaviour.

Operant conditioning is seen in the learning behaviour of many animals, including humans. Oftentimes it goes completely unnoticed, but it can also be used as a deliberate tool in training. As young children explore their world and learn how to eliminate unpleasant stimuli and satisfy desires through proper behaviour, they are learning through operant conditioning.

The Peppered Moth Studies

The Peppered Moth (Biston Betularia)

Within one hundred years (1850 to 1950); the dotted whitish form of the peppered moth (*Biston betularia*) was almost entirely replaced by the melanic (black) form. The melanic form appeared to be best suited for survival against the soot that had collected on forest tree trunks from pollution because the moths could camouflage with their resting area on the tree. The dotted whitish form of the peppered moth was no longer predominant in this environment because they were easily detected and predated on. How does this happen? Many animals have anti-predator adaptations.

Adaptations are defined as "a heritable trait that either spread because of natural selection and has been maintained by selection to the present or is currently spreading relative to alternative traits because of natural selection (Alcock, 2001)." Anti-predator adaptations suggest that a heritable trait, which enables the organism to hide from predators by seeking cover against a background, has spread by natural selection because of reproductive success. H.B. D. Kettlewell's experiments on the peppered moths, as well as, those conducted by R.J. Howlett and M.E.N. Majerus have proven that the peppered moth's preference for their resting places on trees are anti-predator adaptations.

In 1955, H.B. D. Kettlewell published his study on pepper moths: *Selection Experiments on Industrial Melanism in the Lepidoptera.* Kettlewell hypothesized that the dotted whitish form of the peppered moth were more likely to be eaten than the melanic form because they could be easily detected against the soot covered trees. His studies showed that the moths that were easily identified by humans were at a higher risk of predation from birds. The dotted whitish form was at higher risk of predation than the melanic form in the polluted environment.

Howlett and Majerus further examined this hypothesis in their study: *The Understanding of Industrial Melanism in the Peppered Moth,* published in 1987. They tested it by pinning 50 of both forms of the peppered moths on pale and dark tree trunks.

Calculations

Moths in Polluted (dark) Woodland.

Dotted Whitish Form: (30/50) x 100 = 60% Melanic Form: (20/50) x 100 = 40%

60% - 40% = 20% The Dotted Whitish form is predated on 20% more than the melanic form in the polluted woodland.

Moths in Non-polluted (pale) Woodland.

Dotted Whitish Form: (15/50) x 100 = 30% Melanic Form: approx. (30/50) x 100 = 60%

60% - 30% = 30% The Melanic form is predated on 30% more than the dotted whitish form in the nonpolluted woodland.

Their studies showed that the dotted whittish form was preyed on 20% more than the melanic form in the polluted woodland. The melanic form is predated on 30% more than the dotted whitish form in the nonpolluted woodland.

In addition to the difference in predation, their studies also showed that in both the polluted and nonpolluted environments the moths that were located on the limb joints instead of the trunks were less likely to be preyed on.

Howlett and Majerus showed that the dotted whitish form was at a higher risk of predation when resting against the polluted (dark) trees. The melanic form had a greater chance of survival and reproduction because they were less likely to be detected against the soot-covered trees, the same findings of Kettlewell years before. Their experiment proved that the melanic form was an anti-predator adaptation, which is why the dotted whitish form had become so rare. The melanic form had become the dominant trait for survival and reproduction in the polluted woodland.

The Monarch Butterfly

Antipredator Adaptations Used by the Monarch Butterfly

The relationship between prey and predators continually changes. Prey need to find ways to outsmart predators in order to survive and reproduce.most monarch butterflies do not live very long. that's why they are endangered. One way monarch butterflies increase their fitness is by forming huge groups of up to ten million. As their group size increases, the probability that any one monarch butterfly will be captured decreases. Moreover, the butterflies located in the center of a large group are more likely to survive than those on the outside (Pike, 1999). This is known as the "dilution effect (Alcock, 2001)."

Monarch butterflies that choose to migrate to closed-area overwintering sites are less likely to be attacked by a predator. Also, by reacting as a group to the movement of a predator, monarch butterflies are better able to scare away predators (Pike, 1999). This "mass startle effect" is thought to stun the predators and provide time for the butterfly to escape (true shock and awe indeed).

The aforementioned are collective ways in which the butterfly behaves in order to elude a predator. There are, however, certain individual inherent features that the monarch butterflies possess that increase their probability of avoiding a predator.

Monarch butterflies contain chemicals that are toxic to many predators. Evidently, this makes other similar but harmless species envious. In Batesian mimicry, a palatable species attempts to mimic an unpalatable species in an attempt to increase its own fitness. The monarch butterfly species is one that some Batesian mimics model themselves after perhaps because the monarch butterflies are so successful at avoiding predators (Mappes, 1997). The Batesian mimics, although they are not harmful to predators, experience increased fitness because they model a potentially harmful species such as the monarch butterfly. The viceroy butterfly (*Limenitis archippus*) is one such butterfly that models itself after the monarch (Pike, 1999).

Although monarch butterflies do not gain an increase in fitness as models of Batesian mimics, they do benefit from Müllerian mimicry. In Müllerian mimicry, two unrelated, toxic species converge on a similar morphology. If more than one unpalatable species has a similar morphological trait, then the predators may more easily recognise a Müllerian mimic as potentially harmful. This saves both the mimics and the predators time and energy.

Lastly, monarch butterflies display aposematism or warning colouration. This warning colouration is meant to be very conspicuous (Alcock, 2001). Monarch butterflies make themselves conspicuous by having bright orange areas on its wings. Predators quickly learn that prey containing these bright colours are potentially harmful. For example, when a blue jay consumes a monarch butterfly, it vomits shortly after. From that point on, the blue jay associates features of the monarch butterfly, such as its bright colours, as unpalatable (Alcock, 2001). Because of their morphological features, Batesian mimics, Müllerian mimics, and many other aposematic species all gain from the monarch's unpalatability.

Reproductive Strategies in Fish

Reproduction is the sole method through which a species may continue, and reproduction manifests itself in numerous deviations throughout nature. The ultimate goal of each species is to produce the maximum amount of offspring while exerting the least amount of energy. Predation pressures, resource accessibility, and competition with the species for attracting mates greatly influence the likelihood of an individual within a species producing viable offspring and donating genes to subsequent generations. Different fish species may be observed in order to learn the unique reproductive strategies employed throughout nature from fertilization tactics to parental care methods that have adapted these fish in continuing their species.

Females typically exert much more energy and time into reproduction than males, and therefore, females are usually more selective in choosing mates. Females typically favourmales that are larger in size, possess more elaborate physical traits than the other males, and display more energy in courtship activity (http://cetus.ucsd.edu/people/Ana/femalechoice.htm). However, smaller and less desirable males must still be able to fertilize the eggs, or, according to natural selection, they would have been eliminated from the population. Although slightly more unsettling, sexual coercion results in the fertilizing of eggs just as sexual courtship does. The term "sneakers" refers to smaller males who opt to race into the nest more desired males and fertilize the eggs while the nesting male pursues the female (http://users.ipfw.edu/gillespi/BIOL%20598/Articles/Goncalves,%20et%20al.%202003.pdf). The term "satellites" describes cryptic males mimicking females in appearance and behaviour so that they are able to fertilize the eggs while the unsuspecting couple spawn.

Although fish species exhibit both internal and external fertilization, fish also exemplify many unique strategies for parental care. Most fish species utilise broadcast spawning, which is a method a releasing gametes into the water and providing no parental care. The goal of this reproductive method is to produce the maximum amount of progeny in hopes that as many offspring as possible will survive. Many fish species, however, opt to employ parental care in order to supply their smaller number of offspring a greater chance of survival. Oral brooding and sex role reversal are two unique methods of parental care that have proved successful in certain fish species.

Oral brooding, although quite rare in nature, is a parenting method adopted by many fish in the Family Cichlidae (http://www.cichlidae.com/articles/a133.php). After the successful fertilization of the eggs, Cichlids, such as Tilapia, place the egg clutch into the mouth of the female. The fecundity is significantly lower in oral brooders; however, the eggs tend to be larger and receive more nutrients. By placing the eggs into the oral cavity, the female is able to provide protection for the eggs, and the churning ability of the female rotates the eggs so that each egg is exposed to oxygenated water.

Similar to oral brooding, sex role reversal in fish is rare but is utilised in the Family Sygnathildae. In sea horses, the female inseminates the male by inserting the oviduct into the male brooding pouch several times to ensure fertilization. After fertilization is complete, the female departs, and the male attaches itself to a nearby object with its tail waiting for the eggs to mature.

Nature exhibits many alternatives that species have adapted in order to reproduce. By utilising these methods, these organisms are able to reproduce viable offspring and continue their kind. Survival of the species is the ultimate goal for all organisms, and these organisms are constantly adapting to their environment in order to do so.

Polyandry

Polyandry is defined as "the mating of one female with more than one male while each male mates with only one female." Exclusive polyandry (as opposed to polyandry in concert with polygyny) is very rare, occurring in only about 1% of animal populations, most being shorebirds like the sandpiper. The basis of polyandry is a sex role reversal. The females compete for the males and are larger and more colourful, while the males take on the parental role. With the sex role reversal, a natural selection against older males evolves. This is accomplished by the females tending to select the males with the best sperm in order to give the female the most offspring possible. Younger males will more likely have fertile sperm; therefore impregnating the female on more instances than an older male with less fertile sperm.

Types

There are two major types of polyandry: simultaneous and sequential. Simultaneous polyandry is when the female controls a very large territory. Inside this territory, the female has multiple smaller nesting territories with different males. The female mates with all males simultaneously, keeping control of the smaller territories. Another form of simultaneous polyandry, cooperative simultaneous polyandry, is when the female only has one nesting area where she mates with multiple males producing a clutch of eggs of mixed parentage with all males contributing to the eggs. Sequential polyandry, the most common form, is where the female mates and produces a clutch of eggs with one male, then leaves the male to incubate and rear the eggs while moving on to another male in a different nesting territory. Here, the female moves from one male to another, leaving the male in full responsibility of the eggs instead of sharing the responsibility.

Benefits

There are many genetic benefits of polyandry. The first being fertility insurance. This hypothesis suggests that by mating with multiple males, the female is guaranteed to fertilize all of her eggs. The multiple partners potentially make up for one male that may not be able to fertilize the eggs. The good genes hypothesis states that the females have multiple mates because she is in search of the male that will pass along the best genes to her offspring. By finding this male, the female is increasing the survival rate of her offspring. The genetic compatibility hypothesis is one that suggests that the female finds multiple mates in order to find the most compatible genetic match for her eggs. While looking for a good match, she is also eliminating the males that are least compatible with her eggs.

The material benefits of polyandry can be seen through three hypotheses. The more resource hypothesis suggests that the more mates the female has, the more males she has to care for her clutch. The better protection hypothesis states that by having multiple partners, the female is better protected from predators. The infanticide reduction hypothesis is one that claims that since the female has multiple males, she has a lower infanticide rate because the males do not know which progeny belong to them. This prevents the males from killing other male's young. Although some infanticide occurs between females and other female's eggs, it is minimal among males.

Eusociality

Sociality and Eusociality

Social behaviour: behaviour that relates to interactions with other members of the population. Pros/cons of group living. <Society> a group of individuals of the same species that is organised in a cooperative manner extending beyond sexual and parental care.

Group Living, altruism, coorperative behaviours may emerge if genes for such characteristics are favored. This may occur under different circumstances

Egoistic Solutions: Selfish herd; Mutualism although the behaviour may seem altruistic, individuals benefit directly (e.g. living in a group and co-operating with others directly benefits the individual because of the general advantages of group living). Reciprocal altruism where the benefits accrued to the individual exceed the costs it incurs

Kin selection when behaviour sufficiently enhances the fitness of related individuals (i.e., carrying similar genes) leading to an increased representation of genes being passed on to the next generation Eusociality: an extreme form of altruism where sterile workers labouron behalf of reproductive individuals, cooperation in the care of young. There must be an overlap of at least two generations of individuals which share in colony labor.

Inclusive Fitness: refers to the sum of an individual's <Direct Fitness> (probability of reproductive success of one's own offspring) and its <Indirect Fitness> (probability of reproductive success of non-descendant relatives). Altruistic acts are frequently directed towards relatives. <Kin Selection>: selection for traits that lower an individual's personal fitness, but raise a relative's fitness. Recipient (related) kin share genes with the altruistic individual and are thus genetic extensions of them. Kin selection and inclusive fitness can explain sociality through increased survivorship of relatives, increased inclusive fitness, altruism, and delayed maturation or breeding.

Hamilton's Rule benefit to recipient / cost to altruist > 1 / coefficient of relatedness.

Memory

Learning and Memory

Memory, in its broadest sense, refers to an individual's ability to retain learned information. The formation of memory traces is a complex task which all organisms appear to be able to do. Fundamentally, such an ability requires a series of distinct steps: Information must be encoded as meaningful associations are assessed, the resulting information must be recorded in some form, and the memory record must be retrieved when needed. As humans exhibit little conscious awareness of mental processes, such as memory, these were long considered inaccessible to objective, scientific study. Inferences about the course of encoding, consolidation and storage of memory thus mostly rely on indirect methods such as testing for retrieval abilities. Consolidation of memory appears to involve independent paths at different time frames. Such evidence derives from the existence of different critical time frames for its formation, and its susceptibility to disruption with different pharmacological tools.

Sensory Memory

Subjects are able to report great detail about a complex stimulus immedeately following its presentation. This ability forms within a few tens of milliseconds and decays again rapidly within a few hundred milliseconds.

Short-term Memory

Short-term memory allows the recall of something from several seconds to as long as a minute. Its strength does appear to depend primarily on attention and not rehearsal. It is thus highly vulnerable to disruption when attention shifts elsewhere. The amount of information that can be held is on the order of 4-6 numbers. This amount can be boosted by grouping them into distinct chunks as for telephone numbers (e.g., 202-456-1414).

The ability to recall such information is contingent on transient patterns of neuronal activity in regions of the frontal and the parietal lobe.

Long-term Memory

This type of memory, lasting hours to months, critically depends on a transfer of the information from short term memory using repeated rehearsal.

The hippocampus appears to be an essential structure in such routing. Sleep is thought to improve the consolidation of information, possibly by hippocampal replaying of activity from the previous wake period. Electrical activation of hippocampal circuits reportedly are linked to feelings of Deja Vu.

Declarative Memory

Declarative memory refers to the ability to become conscious of, or declare, facts and experiences. It is also referred to as Explicit memory, when it involves direct recalling information that had been obtained from the external world. Representing knowledge of standard textbook material or events, it is best formed by actively recalling the material in spaced intervals. Compared to other forms or memory it is more volative; more easily formed and more easily forgotten.

The primary neural basis appears to reside in the medial temporal lobe. Bilateral damage to this area results in anterograde amnesia, as in the famous case of the anonymus memory-impaired patient HM. A surgical procedure for epilepsy left him with damage to his brain in the medial temporal lobe on both sides. the hippocampal formation, parahippocampal gyrus, the entorhinal cortex, and the amygdala. He subsequently suffered from severe anterograde amnesia, where transfer of new events into long-term memory was impaired. He was unable to recall events once his attention had focused elsewhere.

Semantic Memory is the ability to consciously recall knowledge of facts that are independent of a specific time and place. medial temporal lobe, diencephalon

Episodic Memory refers to the ability to explicitly recall information about a specific event that has occurred at a specific time and place, medial temporal lobe, diencephalon

Nondeclarative Memory

Implicit memory, which does not need to involve conscious awareness in the act of recollection

Procedural Memory regards the learning of motor skills and habits. Formation requires repeated reinforcement, repetition and practice over many trials rather than recollection. Once formed it is less likely to be forgotten. It is also less easily transferred to related tasks than declarative learning

Striatum, basal ganglia, Deficits can be assessed using a serial reaction time (SRT) task, backwards reading, mirror drawing, probabilistic classification, artificial grammar learning, or prototype abstraction.

Motor Responses with Classical Conditioning, Cerebellum.

Emotional Responses with Fear conditioning involves an organism's ability to acquire fear responses to a previously neutral stimulus. This occurs when it becomes paired with an aversive stimulus, such as a shock or loud noise. James Wattson's Little Albert experiment illustrated that children learned fear of objects when encounters with them where paired with loud noises.

Priming refers to a context-dependent activations of clusters of neocortical neurons. As they become more activated, they are more likely to come into consciousness.

Even reflex pathways are capable of surprising plasticity. For example, circuits that control motor patterns for walking will be subjected to optimisation with sensory input.

Long-lasting Memory

Months to lifetime.

Case Studies

Patient HM suffered severe memory impairment following bilateral damage to his medial temporal lobes, hippocampal formation, parahippocampal gyrus, entorhinal cortex, and amygdala. His short-term memory remained intact, he was able to largely recall past events, and his ability to learn new motor skills was not disrupted. However, he suffered from severe anterograde amnesia, i.e., the ability to commit new events to long-term memory.

Mechanisms

Memory mechainsms critically depend on changes in synaptic functioning. Hebb Synapses and their role in classical conditioning. If a synapse succeeds at driving a postsynaptic neuron above threshold, its subsequent effectiveness is strengthened.

Long-term potentiation is based on changes in neural signals which will potentiate a neural response for 1-2 weeks. potentiation involves increased release of the excitatory neurotransmitter glutamate.

Molecular Changes at the Synaptic Level

During habituation, repeated stimulation of a sensory neuron leads to a smaller activation of the postsynaptic motor neuron. The primary cause for this is a progressive reduction of Ca inflow into the presynaptic terminal, decreased transmitter release, and a smaller activation of the postsynaptic target.

In sensitization as a result of electrical head shock, release of serotonin activates second messenger systems and phosphorylation of key target molecules. Long-Term Memory depends on new protein synthesis and the formation of new synaptic connections.

Neurophysiology

Neurons

The basic functional unit of the nervous system are neurons, i.e., cells that are specialised for processing and communicating information. Neurons have a basic structure of:

- The cell body (or soma) is the bulbous end of a neuron, containing the cell nucleus and much of the cells metabolic machinery.
- The axon carries information away from the cell body and may range in size from several microns to as long as several meters in giraffes and whales. Axons may branch into terminal buttons (designated sites of communication) at its end.
- Dendrites receive neurotransmitter secreted by the axon of other neurons.
- Synapse is a specialised junction between cells of the nervous system which permits signalling between them.

Neurophysiology

Neurons are able to conduct and process signals that involve small changes in electrical voltage generated across the neuronal membrane. At rest, the inside of the neuron is negative relative to the cell surround - the resting membrane potential.

Individual signals are integrated across different regions of the neuron and throughout sets of neuronal networks in order to determine each unit's response and output. A special type of such voltage changes features a rapid spike in voltage resulting from changes in ion conductances - the action potential. Such signals are predominantly transmitted along axonal lines where neurons interact with other cellular entities over greater distances.

Ohm's Law

Basic familiarity with key concepts in electricity is essential in order to appreciate the neuronal mechanisms active during such integration. Ohm's law holds that the rate at which electric charges flow (i.e., Current or I) depends on the force and its direction exerted onto the charged particle (i.e., Potential, V) and the ease with which the flow can occur (i.e., Conductance, g). The latter can also be expressed as the reciprocal of the conductance, namely as the degree to which the conductor obstructs the flow of charges (i.e., Resistance, R). The flow of electricity is defined as flowing from negative to positive.

Ohm's Law: I = g * V

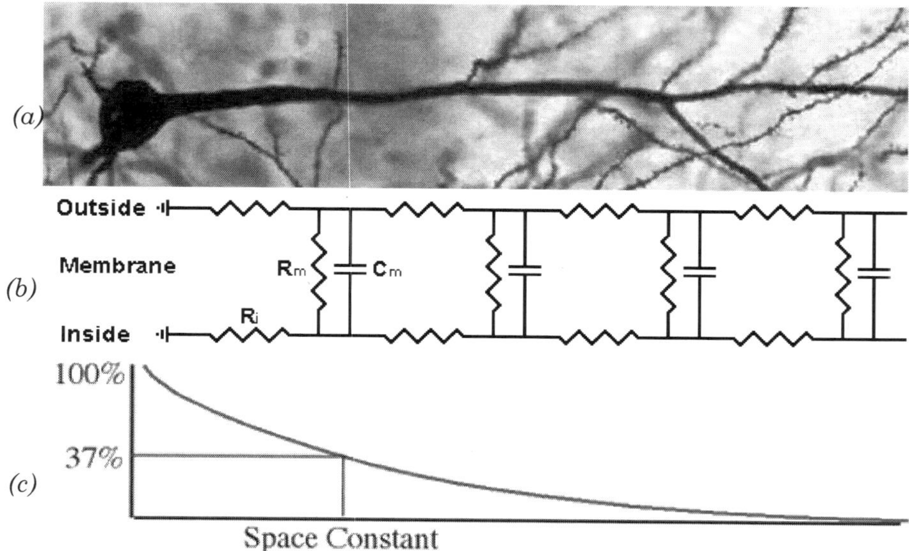

Figure: *The electric properties of the neuronal membrane can be expressed as a circuit diagram. a. Golgi-stained neuron with long axonal process. b. Circuit diagram of neuronal membrane combining membrane resistance (Rm), membrane capacitance (Cm) and internal or axial resistance (Ri). c. Decay in signal size over distance with Space Constant. d. Change along the axon in Voltage response to the injection of a single square current pulse due to membrane capacitance*

Electrical Properties of Neurons

The voltage present across the cell's membrane can be measured between a reference electrode and a conducting glass capillary tube drawn out to a very fine point, and inserted into the cell. At rest the neuronal interior is negative relative to the outside at between -60 to -80mV, while most non-neuronal cells feature a potential of about -30 mV.

The current is carried by ions moving across the membrane at ion conductances when specific channels allow them to pass. Ignoring its active properties, the axon can in electrical terms be viewed as an insulated cable. An electric potential is able to spread passively along any stretch of membrane. In the process its strength decays and the slope of on- and offsets becomes less steep with distance due to the passive cable properties of the membrane. The latter include resistance along its length and both a resistance and a capacitance component across it. Signals will spread fastest when longitudinal resistance is low (e.g., via increased axonal diameter) and they will spread furthest when resistance across the membrane is high (e.g., with layers of myelin for added electrical insulation).

Resting Potential

A neuron is at rest when it has not been activated for some time. At that state it exhibits a negatively charged inside relative to the surrounding encironment.

Ionic Basis

The Resting Potential emerges from an unequal distribution of ions across the cell's membrane. An excess of K^+ and of various $Anions^-$ exists on the inside while Na^+ and Cl^- exist in higher concentrations on the outside. The membrane is dotted with proteins that permit the flow of ions (i.e., ion pores) or control their flow (i.e., ion channels) across it. In a cell at rest only potassium ions (K^+) are able to flow freely between the two compartments. Initially we observe a net flow of positively charged potassium ions from the inside where they are more numerous to the outside where their concentration is much lower (i.e., concentration gradient). With a net flow of positively charged ions from the inside across the membrane, the inside becomes increasingly negative relative to the outside (i.e., charge gradient).

The resting potential settles into an equilibrium when as many potassium ions are pushed out of the cell along their concentration gradient, as will enter the cell along the accompanying electrical gradient. This equilibrium potential for potassium can be calculated using the Nernst potential for known inside and outside concentrations of any given ion

Nernst Potential (E) = $2.303 * {}^{RT}/_{zF} * \log({}^{[ion]out}/_{[ion]in})$

where R = the gas constant (8.3143 joules/mole-degree); T = absolute temperature in degrees Kelvin (310 degrees); z = ionic valence; F = Faraday's constant (96,487 coulombs/mole).

Depending on their respective inside and outside concentrations at the particular cell, different ions will produce different equilibrium potentials (reversal potentials). The ion concentrations are usually similar to those listed:

Ion	$[ion]_{in}$	$[ion]_{out}$	***E***
K^+	400	20	-75mV
Na^+	50	440	+55mV
Ca^{2+}	0.0001	125	+155mV
Cl^-	9	100	-65mV

The Nernst potential for potassium thus can be simplified to the following formula. $E_k = 58mV \times \log\frac{20}{400} = -75mV$ and for sodium is:

$$E_{Na} = 58mV \times \log_{10}\frac{440}{50} = +55mV$$

The cell's resting membrane potential combines all relevant ion currents with K^+ ions figuring most prominently due to their high resting conductance. In addition a small number of Na^+ ions leak into the cell.

The resulting resting membrane potential is thus slightly lower compared to the E_K+ at around -65mV. To maintain the concentration gradient across the membrane of resting neurons, ATP continually supplies energy to the Na^+/K^+ ATPase (ion pump) as ions leak across the axomembrane.

Action Potential

An Action potential represents a transient inward current of cations (Na^+). It is generated when the graded membrane potential rises above threshold either spontaneously or because of depolarising input.

Ionic Basis

During the action potential, voltage-gated sodium channels are opened when the potential rises (i.e., depolarises) from resting potential (-65mV) to the cells threshold (-55mV). These channels close again when the potential further rises to +10mV. As sodium ions spread to neighboring areas of axonal membrane, they open sodium channels there via depolarisation and the signal runs the entire length of the cell (i.e., the *all-or-none-principle*).

A Refractory period follows the action potential during which time the neuron reestablishes its normal resting potential. At the beginning of this period it is impossible for another signal to be transmitted, this is called *absolute refractory phase.* This is followed by the *relative refractory phase* where it is possible to send another signal but more exitation than normal is needed.

The Duration of an action potential is 1-2ms in Vertebrates and 1-100ms in Invertebrates. Frequency of firing ranges from <1 to about 100/sec (100Hz). The Amplitude ranges between 70-80mV when recorded intracellularly and 5-200μV when recorded extracellularly.

Synapse

Electrotonic Junctions: Two electrically excitable cells, such as neurons or muscle cells, may be electrically coupled where an action potential in one cell moves directly into the other via arrays of gap junctions. Electrical synapses are fast but cannot be modulated. They are mostly used in neuronal circuits for escape behaviours where speed of conduction is essential. Electrical synapses (gap junctions, electrotonic junctions) allow current to flow between separate neurons when ions pass through gap junctions. Connexons (where 6 connexin proteins form a hemi-channel) are the actual pores that allow ions to flow past the two membranes. A connexon in the presynaptic membrane lines up precisely with its respective equivalent in the postsynaptic membrane, forming a continuous channel from one neuron to another. With a pore diameter of about 1.5m-9 many small molecules can pass through efficiently. Intracellular Ca2+ concentration, pH, or phosphorylation of connexins can profoundly alter the easy with which ions annd proteins may pass through the pore. As there is no synaptic delay in transmission of current from cell to another, the conduction of potential changes is considerably faster than through chemical synapses. Although electrical synapses are often bi-directional, some synapses pass current better in one direction than the other (i.e., rectifying synapse) Electrical synapses are commonly used in time-critical processes (escape behaviours), when rapid synchronisation of many cells is needed (e.g., vertebrate cardiac muscle), between glial cells, or early in development.

Chemical Junctions

Cells are also able to communicate chemically across a gap (i.e., synapse), which forms a directional connection from one neuron to another cell. The neurotransmitter is emitted from the terminal endings of one cell's axon onto a dendrite or body of a second cell. The arrival of a signal leads to the release of a neurotransmitter from the presynaptic terminal, diffusion across the synaptic cleft, and binding to receptors in the postsynaptic membrane. Ionotropic or metabotropic receptors are binding proteins which alter ion conductances at the postsynaptic cell membrane (e.g., increase in Na^{+} conductance depolarises and is excitatory, CL^{-} hyperpolarises and is thus inhibitory).

When excitatory, the postsynaptic cell is depolarised with an excitatory post-synaptic potential (EPSP) and an action potential is elicited if the threshold is reached; in inhibitory connections, the postsynaptic cell is hyperpolarised with an inhibitory post-synaptic potential (IPSP) and it will thus be harder for other inputs to drive the cell towards an action potential. A single input is rarely sufficient to lead to an action potential in the post-synaptic cell. Multiple EPSPs may add and reach the threshold when a series of action potentials arrive at high rate. Chemical synapses are capable of integrating a complex scenario of inputs:

Neurotransmitter refers to a compound that is released at a synapse and diffuses across the synaptic cleft to act on a receptor located on the membrane of a postsynaptic cell, which may be another neurone, a muscle cell or a specialised gland cell. A wide range of chemicals are used as neurotransmitters in the nervous system. Stored in synaptic vesicles. they are released during the arrival of an action potential and produce changes in the excitability of the postsynaptic membrane. Ca^{2+} influx at the axon terminus is required for synaptic release. Neuromodulator refers to a compound that is released within a localised region of CNS, the receptor for which is not necessarily sited on an anatomically apposed postsynaptic cell. Thus a neuromodulator may affect several postsynaptic cells with specificity conferred mainly by the distribution of receptors. Main action is on second messenger systems, e.g. cAMP or inositole triphosphate, presumably affecting protein phosphorylation

Sometimes the same neurochemical may have rapid transmitter type effects, followed by longer modulatory influences. This suggests that neurotransmitter and neuromodulator effects may be most effectively classified at the receptor level. Activation of receptors on a protein structure directly incorporating an ion channel (a ionophore) are defined as neurotransmission while activation of receptors coupled indirectly to ion channels (e.g. via second messenger systems) are defined as neuromodulation.

Signal Strength

For the signal to be passed from one neuron to the next it must have enough energy to break a point called the threshold. Once the threshold is broken the signal is transmitted. The neuron fires at the same strength every time. The strength of a signal is decided by how many different neurons are being fired and at what frequency they are being fired.

Glial cells

The amount of' glial cells to every neuron in the nervous system is disputed. Glial cells function as support for the neurons; they produce the myelin sheath which surrounds some neurons and also form part of the *blood-brain* barrier. The blood-brain barrier is a structure that prevents certain substances in the bloodstream from reaching the brain. Many axons are sheathed with tubes of myelin, which is a fatty material. Myelin is produced by the *glial* cells. The myelin sheaths on axons have gaps, which are called the nodes of Ranvier. Myelinated sheaths help transmit information quickly and efficiently.

Synaptic Integration

Activity in any presynaptic neuron alone is rarely sufficient to elicit an action potential in the postsynaptic neuron. Summation of synaptic inputs, however, can occur when a neuron receives multiple exitatory inputs in short order. Synaptic processing describes the ways in which the electrical events in different neurons may interact through interactions at their synapses. Synaptic inputs are mostly located on the dendritic tree of a neuron, while axonal action potentials represent the neuron's output. In order to drive a neuron the integrated influence of the electrical events generated by the synapses must be sufficient to spread through the dendrites to the soma and into the initial segment of the axon (i.e., the axon hillock, trigger zone). The latter is the site of action potential generation. Synaptic integration is thus key to understanding the contribution of individual neurons to information processing within neural networks.

In spatial summation the cell responds when activity from multiple inputs arrives at the same time. If the combined excitation brings the cell to threshold, an actionpotential is initiated. The arrival of a spike train in a presynaptic neuron may produce multiple EPSPs. If these inputs begin to build on each other, the postsynaptic cell may reach threshold and produce an action potential. Passive membrane properties are important. Long time constants increase the chances for temporal summation. Large space constants determine the likelihood of spatial integration. The size of the postsynaptic cell counts, as a synaptic current delivered into a small neuron will produce a much greater effect than in a large postsynaptic cell.

Location of synapses is important. Control of the cell's activity usually driven by excitatory synapses on dendrites. Inhibitory synapses located on the cell body are able to shut down activity. Synapses at presynaptic release sites are modulatorym controlling the amount of neurotransmitters released.

Chapter 2

Nesting Behaviour in Animal

Over the course of a turtle's nesting season, a female sea turtle will leave her homing ground and migrate to nesting grounds to build a nest and lay eggs. One species of turtle, *Chelonia mydas*, the Green sea turtle, lives off the coasts of Brazil but nests each year on the beaches of Ascension Island, a tiny island in the mid-Atlantic Ocean (Papi et al.). These turtles typically make their way to the island between the months of December and March (Sims).

After migrating to a nesting ground, it is time for the female to choose a nesting site along the beach. Many factors play into this decision and a female will scope out a site for a few days before she actually decides to build a nest. However, once the decision of where to nest has been made, females will most likely nest at the same beach and site for the rest of their reproductive years (CCC&STSL).

All species of sea turtles exhibit similar nesting behaviour, possibly differing only in inter-nesting period length (the period of time between two nestings during the same nesting season). On the night of nesting, a sea turtle will emerge from the sea and scope out her surroundings one more time before digging a nest.

If there seems to be any unnatural distractions (i.e. lighting or human presence), a female may return to the sea without building a nest ("false crawl"). Most often, however, she will proceed to dig a nest on the beach. To do this, the female turtle first digs a relatively shallow depression, known as a body pit, in the sand using her front flippers and body to push sand out of the way. Then she will excavate a single egg chamber, a deeper cavity opening to the body pit, into which she will oviposit an entire clutch of eggs during nesting. Finally, after constructing this egg chamber, she will begin ovipositing her eggs. It is extremely rare that a sea turtle will stop nesting before laying is complete but will happen if the female senses extreme danger (CCC&STSL).

After laying eggs, the female will camouflage her nest by filling in the body pit with sand and return to the sea only to nest at the same site at a later date in the nesting season. After nesting a second time, the female sea turtle will return to the sea and migrate back to her homing ground (CCC&STSL).

Many nesting grounds, such as that of the Green sea turtle, are extremely far from the species' homing ground. This migration of sea turtles during the nesting season has led to the development of many different hypotheses of *how* these adult female turtles continually find the common nesting ground and then how offspring navigate to the homing grounds. First, it has been suggested that sea turtles may use the earth's magnetic field to aid in navigation between homing grounds and nesting grounds.

There have, however, been experiments that have weakened this hypothesis in adult sea turtles. One such experiment involved placing magnets on the back of experimental Green sea turtles in order to disrupt the magnetic field possibly detected by the sea turtles and the courses these turtles followed on the return journey from Ascension Island were tracked. The results show that, although hatchlings may be able to navigate via earth's magnetic field, it is very possible that this ability is absent in adults (Papi et al.). However, another hypothesis has been suggested and states that turtles may use olfactory cues from the ocean currents and winds to navigate through the ocean to nesting grounds (Sims).

Maturation

Individuals, during their progression from infancy to adulthood, experience a variety of changes, including physical, cognitive, behavioural, and psychosocial.

Puberty

Successfully navigating the transition from juvenile, to meet the demands of adulthood, remains one of the trickiest tasks any individual will likely face. The changes in behavioural repertoire that occur during puberty are reflected in the maturation of endocrine, neural and neurochemical systems. The influence of rising gonadal hormones on male and female reproductive behaviours are well understood.

However, corresponding changes also exist in structure and function of neural circuits underlying a much broader range of behavioural and cognitive development.

Meerkat

The meerkat or suricate, *Suricata suricatta,* is a small mammal belonging to the mongoose family. Meerkats live in all parts of the Kalahari Desert in Botswana, in much of the Namib Desert in Namibia and southwestern Angola, and in South Africa. A group of meerkats is called a "mob", "gang" or "clan". A meerkat clan often contains about 20 meerkats, but some super-families have 50 or more members. Meerkats have an average life span of 12–14 years.

Name

"Meerkat" is a loanword from Afrikaans. The name has a Dutch origin but by misidentification. Dutch *meerkat* refers to the "guenon", a monkey of the *Cercopithecus* genus. The word "*meerkat*" is Dutch for "lake cat", but the suricata is not in the cat family, and neither suricatas nor guenons are attracted to lakes; the word possibly started as a Dutch adaptation of a derivative of Sanskrit *markaþa*. The traders of the Dutch East India Company were likely familiar with monkeys, but the Dutch settlers attached the name to the wrong animal at the Cape. The suricata is called *stokstaartje* = "little stick-tail" in Dutch.

According to African popular belief (mainly in the Zambian/ Zimbabwean region), the meerkat is also known as the sun angel, as it protects villages from the moon devil or the werewolf which is believed to attack stray cattle or lone tribesmen.

Anatomy

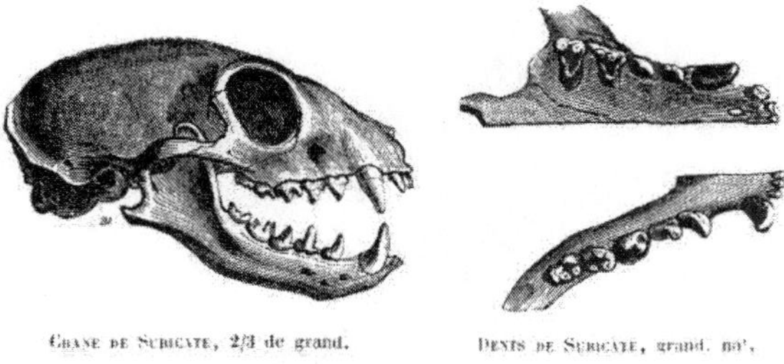

Figure: *Skull and dentition, as illustrated in Gervais'* Histoire naturelle des mammifères

The meerkat is a small diurnal herpestid (mongoose) weighing on average about 731 grams (1.61 lb) for males and 720 grams (1.6 lb) for females. Its long slender body and limbs give it a body length of 25 to

35 centimetres (9.8 to 14 in) and an added tail length of 17 to 25 centimetres (6.7 to 9.8 in). Its tail is not bushy like all other mongoose species, but is rather long and thin and tapers to a black or reddish coloured pointed tip.

The meerkat uses its tail to balance when standing upright. They also use it for signal. Its face tapers, coming to a point at the nose, which is brown. The eyes always have black patches around them and it has small black crescent-shaped ears that can close to exclude soil when digging. Like cats, meerkats have binocular vision, a large peripheral range, depth perception, and eyes on the front of their faces.

At the end of each of a meerkat's "fingers" is a non-retractable, strong, 2 centimetres (0.79 in) long, curved claw used for digging burrows and digging for prey. Claws are also used with muscular hindlegs to help climb trees. They have four toes on each foot and long slender limbs. The coat is usually fawn-coloured peppered with gray, tan, or brown with a silver tint. They have short parallel stripes across their backs, extending from the base of the tail to the shoulders. The patterns of stripes are unique to each meerkat. The underside of the meerkat has no markings, but the belly has a patch which is only sparsely covered with hair and shows the black skin underneath. The meerkat uses this area to absorb heat while standing on its rear legs, usually early in the morning after cold desert nights.

Diet and Foraging Behaviour

Figure: *Meerkat killing an elephant shrew, as illustrated in Brehm's* Life of Animals

Figure: *Meerkats showing foraging behaviour at the Adelaide Zoo.*

Meerkats are primarily insectivores, but also eat lizards, snakes, scorpions, spiders, plants, eggs, small mammals, millipedes, centipedes and, more rarely, small birds. Meerkats are immune to certain types of venom; they are immune to the very strong venom of the scorpions of the Kalahari Desert, unlike humans. They have no excess body fat stores, so foraging for food is a daily need.

Meerkats forage in a group with one "sentry" on guard watching for predators while the others search for food. Sentry duty is usually approximately an hour long.A meerkat can dig through a quantity of sand equal to its own weight in just seconds. Baby meerkats do not start foraging for food until they are about 1 month old, and do so by following an older member of the group who acts as the pup's tutor. The meerkat standing guard makes peeping sounds when all is well. If the meerkat spots danger, it barks loudly or whistles.

Reproduction

Meerkats become sexually mature at about one year of age and can have one to five pups in a litter, with three pups being the most common litter size. Wild meerkats may have up to four litters per year. Meerkats are iteroparous and can reproduce any time of the year but most births occur in the warmer seasons. The pups are allowed to leave the burrow at three weeks old. When the pups are ready to emerge from the burrow, the whole clan of meerkats will stand around the burrow to watch. Some of the adolescents might try to show off so they can have more attention than the pups.

There is no precopulatory display; the male ritually grooms the female until she submits to him and copulation begins, the male generally adopting a seated position during the act. Gestation lasts approximately 11 weeks and the young are born within the underground burrow and are altricial (undeveloped). The young's ears open at about

15 days of age, and their eyes at 10–14 days. They are weaned around 49 to 63 days. They do not come above ground until at least 21 days of age and stay with babysitters near the burrow. After another week or so, they join the adults on a foraging party.

Usually, the alpha pair reserves the right to mate and normally kills any young not its own, to ensure that its offspring has the best chance of survival. The dominant couple may also evict, or kick out the mothers of the offending offspring.

New meerkat groups are often formed by evicted females pairing with roving males.

If the members of the alpha group are relatives (this tends to happen when the alpha female dies and is succeeded by a daughter), they do not mate with each other and reproduction is by group females stray-mating with roving males from other groups; in this situation, pregnant females tend to kill and eat any pups born to other females.

Behaviour

Meerkats are small burrowing animals, living in large underground networks with multiple entrances which they leave only during the day. They are very social, living in colonies averaging 20–30 members. Animals in the same group regularly groom each other to strengthen social bonds. The alpha pair often scent-mark subordinates of the group to express their authority, and this is usually followed by the subordinates grooming the alphas and licking their faces. This behaviour is also usually practiced when group members are reunited after a short period apart. Most meerkats in a group are all siblings or offspring of the alpha pair.

Meerkats demonstrate altruistic behaviour within their colonies; one or more meerkats stand sentry while others are foraging or playing, to warn them of approaching dangers. When a predator is spotted, the meerkat performing as sentry gives a warning bark, and other members of the gang will run and hide in one of the many bolt holes they have spread across their territory. The sentry meerkat is the first to reappear from the burrow and search for predators, constantly barking to keep the others underground. If there is no threat, the sentry meerkat stops signaling and the others feel safe to emerge.

Meerkats also babysit the young in the group. Females that have never produced offspring of their own often lactate to feed the alpha pair's young, while the alpha female is away with the rest of the group. They also protect the young from threats, often endangering their own lives. On warning of danger, the babysitter takes the young underground to safety and is prepared to defend them if the danger follows. If retreating underground is not possible, she collects all young together and lies on top of them.

Meerkats are also known to share their burrow with the Yellow Mongoose and ground squirrel, species with which they do not compete for resources. If they are unlucky, sometimes they share their burrow with snakes. Like many species, meerkat young learn by observing and mimicking adult behaviour though adults also engage in active instruction. For example, meerkat adults teach their pups how to eat a venomous scorpion: they will remove the stinger and help the pup learn how to handle the creature.

Despite this altruistic behaviour, meerkats sometimes kill young members of their group. Subordinate meerkats have been seen killing the offspring of more senior members in order to improve their own offspring's position. Meerkats have been known to engage in social activities, including what appear to be wrestling matches and foot races.

Vocalisation

Meerkat calls may carry specific meanings, with particular calls indicating the type or the location of a predator. They chirrup, trill, growl, or bark, according to circumstances. How these calls work is not yet clear.

Meerkat Groups

A meerkat group may die out because of predator attack, its alpha pair being unable to breed, starvation due to drought, or epidemic disease. A new meerkat group often arises from evicted females meeting and staying with roving males, seeking to mate. The litter size is usually 2–5 pups.

The size of the groups is variable. A group which becomes over-large may routinely have to disperse widely to find enough food when foraging. As a result, when suddenly needing to run for shelter, members of the group may choose different holes, resulting in the group fissioning.

Subspecies

There are three subspecies of meerkat:

- *Suricata suricatta siricata*
- *Suricata suricatta majoriae*
- *Suricata suricatta iona.*

In Popular Culture

Meerkats have been the subjects of many cultural depictions, including:

- the Animal Planet series *Meerkat Manor*
- Timon, a character of the Disney animated film's *The Lion King*, The Lion King II: Simba's Pride, The Lion King 1½ and the spin-off series The Lion King's Timon and Pumbaa.
- Aleksandr Orlov, a computer generated meerkat, featured in an advertising campaign, which had to be re-imagined to explain how South African animals come to speak with Russian accents.
- *Life of Pi*, a fantasy adventure novel by Yann Martel published in 2001.

Domestication

Meerkats make very poor pets. They can be aggressive especially toward guests and may bite. They will scent-mark their owner and the house (their "territory").

Antipredator Adaptation

Antipredator adaptations are evolutionary adaptations developed over time, which assist prey organisms in their constant struggle against their predators. The act of a predator acquiring a food source can be divided into four stages: detection, attack, capture and consumption. At every stage in this predatory sequence, adaptations that maximise the prey's chance of survival have evolved. This, in turn, has driven adaptation in their predators. This kind of interaction over long periods is known as co-evolution.

Many predatory animals are themselves preyed upon. To defend themselves, predatory animals often use their methods of attacking prey to inflict or threaten grievous injury to their own predators. For example, the electric eel uses the same electrical currents to kill prey, and to defend itself against predators (such as anacondas, caimans, jaguars, egrets, cougars) which prey on fish of similar size to the eels.

Animal Adaptations

Avoiding Detection

For a predator to locate a potential meal, it must first identify an organism as prey. Prey, however, have many adaptive characteristics which make such a task difficult. Crypsis is the ability of prey to evade detection by predators (or vice versa).

Camouflage is one heavily utilised method, and involves the mimicking of the colour and patterns of the environment. This is achieved by external pigmentation patterns, and resting immobile and silent on an appropriate substrate, as do Egyptian Nightjar. Most marine species possess some degree of countershading, so as to reduce their visibility from above and below. Some, such as the Mimic Octopus, are even able to actively redistribute pigment to provide a better defence. Some animals provide structures on their bodies for plant life to grow, camouflaging its host. An advanced form of camouflage is seen in animals that masquerade as a specific, inedible object: the Bird Dropping Spider is an obvious example.

Figure: *A single zebra is hard to discern amongst a herd*

Some animals have evolved disruptive colouration, whereby instead of attempting to conceal themselves, they instead conceal their body shape, confusing a predator and making identification harder. An example would be zebras, who very much stand out in the savannah when alone, but *en masse*, their stripes allow them to merge together in the eye of a predator into a single confusing creature.

Polymorphism, the existence of many different visual forms within a single species, is a strategy used to reduce predation risk. Predators make use of apostatic selection in order to differentiate edible objects from all other objects; this however means search images must be formed. Experiments have shown that polymorphic prey suffered less predation than single-morph species at a particular density, and polymorphic prey could maintain higher population densities for a given rate of predation.

Animals alter the period in which they are awake in order to avoid predators. Generally, animals are either diurnal, active during the day, nocturnal, active during the night, or crepuscular, active during twilight, depending on food availability, and predator prevalence. Highly mobile creatures such as seabirds migrate to avoid predators during the period in which they are most vulnerable, the breeding season, moving to offshore islands to establish colonies far from the reach of land predators.

Avoiding Attack

Having been detected by a predator, many animals attempt to signal to a predator that they are not worth eating. Some animals make use of bright warning colouration, so as to advertise that they are poisonous. These patterns are often convergent, with red or yellow coupled with black being widely recognised as dangerous. Batesian mimicry is the imitation by a harmless species of the warning signals of a harmful species directed at a common predator, as can be seen in a hoverfly mimicking a wasp.

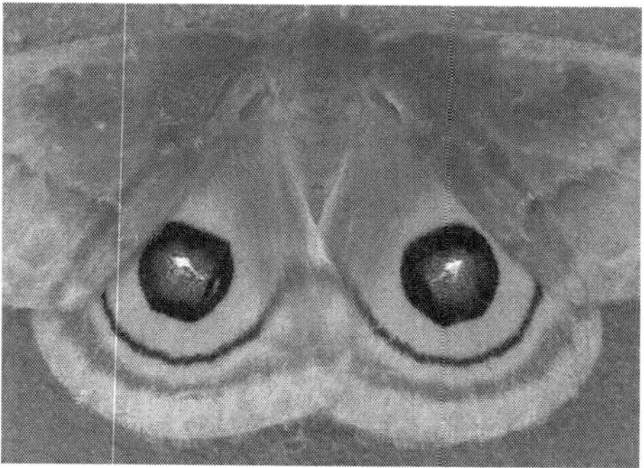

Figure: *The eyespots of* Automeris io

Some animals, particularly gazelles, are known to stot, which, among other things, may advertise their unprofitability to predators. The moth *Automeris io* possesses eyespots hidden on its hindwings. When under threat, the moth suddenly reveals these spots, aiming to startle its predator. Fearful of their own consumption, predators often retreat when startled. Eyespots are also used by some animals to trick a predator into thinking it has been spotted (when in fact, it may not have been), and so is less likely to succeed in pursuit.

Avoiding Capture

Many animals have highly developed senses of sight, smell, and hearing so that they can detect danger and escape. By frequently scanning and monitoring their surroundings, especially when in the open, prey can avoid attack by hoping to see a predator before it reaches the 'critical distance' for an attack. This is a standard defence mechanism for animals in open grasslands and prairies. It is also common for arboreal animals to scan both the ground around them for terrestrial predators, and the sky for aerial predators. Smaller animals may not venture too far from cover in burrows or the undergrowth, where they can quickly hide when danger approaches. Flight is of huge importance in the avoidance of predators in those species that possess it.

Animals that are the frequent target of predation often make use of 'safety in numbers'. This results in a situation where any one herd member is unlikely to be preyed upon, and in high populations, predator satiation is likely to occur. Grazing mammals often feed in social groups, also known as herds. Working as a group, a stalking predator is likely to be detected earlier, and when it attacks, the herd scatters, causing difficulty for the predator and allowing most, if not all, of the prey animals to escape. Prey animals may use alarm signals to alert the other herd members when a predator is sighted or sensed. Animals usually have a breeding season, where all the members of the species spawn at the same time, in order to maximise their young's chance of survival. This is particularly pronounced in insects such as *Magicicada* and the Mayfly, where millions of individuals emerge from pupation on the same day.

Figure: *Muskoxen in defensive formation, horns bared and highly alert*

Many animals use horns, claws, and teeth to fight off predators. Some can inject venom and toxins, and skunks and bombardier beetles spray noxious chemicals to deter attackers. Mobbing, the harassing of a predator by many prey animals, is common in birds, and is usually done to protect the young in social colonies. The Eastern Honeybee mobs invading hornets, vibrating their flight muscles in order to raise the temperature around the hornet scout to lethal levels, rather than allowing the scout to bring others to their beehive.

Avoiding Consumption

Having been captured, an animal must prevent the predator from killing and eating it. Mechanical defences in those that possess them, such as armour and spines prevent access to softer edible parts. Distraction displays are used to direct the attention of the predator away from some vital area, such as the head, or a nest of chicks. These can be visual, as in misdirecting eyespots, behavioural, such as a mother-bird feigning injury, or chemical, as seen in the production of ink clouds by squid and octopuses.

Figure: *This fishing spider has shed two of its legs to escape a predator*

Some animals (e.g. possums) freeze in cover or play dead when seen. Often this is accompanied by foul smells, as if their corpse is in an advanced state of decomposition, one which many animals would avoid consuming. As an absolute last line of defence, many insects taste extremely bitter, so that once a predator has begun to feed they quickly stop, and, if this is a less vital area such as a leg or a wing, the insect survives the encounter.

Some animals take a more drastic approach to defending themselves. Autotomy, the shedding of a non-vital bodypart, is utilised by some species to escape the grasp of a predator. Many lizards shed their tails when clasped, and arthropods will readily give up several legs if allows their escape. When under threat, sea cucumbers rapidly eviscerate, ejecting part or all of their digestive tract. This is done to either anchor the cucumber into a rock fissure, or to eject toxins at the predator. Horned lizards, when threatened, increase the pressure in their sinus cavities until the blood vessels in the corners of the eyes burst, squirting blood at the attacker. Armoured crickets and many other insect species also utilise this method, known as autohaemorrhaging. These animals are able to survive the loss of these tissues, and later regenerate them.

The soldier ant caste of the Malaysian species *Camponotus saundersi* undergoes a process known as autothysis to defend their ant colony. The soldier ants have two large glands that run the entire length of their body, and when stressed during battle, abdominal muscles contract, causing the glands to explode, killing the ant, but spraying poison in all directions.

Ultrasound Avoidance

Ultrasound avoidance is an escape or avoidance reflex displayed by certain animal species that are preyed upon by echolocating predators. Ultrasound avoidance is known for several groups of insects that have independently evolved mechanisms for ultrasonic hearing. Insects have evolved a variety of ultrasound-sensitive ears based upon a vibrating tympanic membrane tuned to sense the bat's echolocating calls. the ultrasonic hearing is coupled to a motor response that causes evasion of the bat during flight.

Although ultrasonic signals are used for echolocation by toothed whales, no known examples of ultrasonic avoidance in their prey have been found to date.

Ultrasonic hearing has evolved multiple times in insects: a total of 19 times. Bats appeared Eocene era, (about 50 million years ago); antibat tactics should have evolved then. Antibat tactics are known in four orders of Insecta : moths (Lepidoptera), crickets (Orthoptera), mantids (Dictyoptera), and green lacewings (Neuroptera). There are hypotheses of ultrasound avoidance being present in Diptera (flies) and Coleoptera (beetles).

Ultrasound Avoidance in Moths

The idea that moths were able to hear the cries of echolocating bats dates back to the late 19th century. White, in a 1877 letter to *Nature* made the association between the moth's high-pitched sounds and the high-pitched bat calls and wondered whether the moths would be able to hear it. However, it was not until the early 1960s that Kenneth Roeder *et al.* made the first electrophysiological recordings of a noctuid moth's auditory nerve and were able to confirm this suspicion). Later research showed that moths responded to ultrasound with evasive movements. Moths, as do crickets and most insects that display bat avoidance behaviours, have tympanic organs that display phonotactic and directional hearing; they fly away from the source of the sound and will only have the diving behaviour considered above when the sound is too loud- or when, in a natural setting, the bat would be presumably too close to simply fly away.

It was found that the moths' responses vary according to ultrasound intensity, diving towards the ground if the pulse was of a high amplitude, or flying directly away from the sound source if the sound amplitude was low (if the sound was softer). Acoustic sensory receptors in noctuid moths are mechanoreceptors located in a chamber formed by the wall of the abdomen and the tympanic membrane, are most sensitive to lower frequencies of ultrasound (between 20–30 kHz.). The moth's body axis allows it to be more sensitive to sounds coming from particular directions. Their ears, on either side of the metathorax, have two sensory cells within the membranes.

Though the tuning curves of these cells are identical, the sensitivity thresholds differ, allowing for sound localisation and a wider range of sensitivity to sound. The movement of the wings during flight also plays a role, since sound thresholds change with wing position. The neural mechanisms for triggering the acoustic startle response are partially understood. However, these is little known about the motor control of flight that ultrasound initiates.

Further research has shown that many species of moths are sensitive to ultrasound. Sensitivities for ultrasound change according to the environment the moth thrives in, and the moth can even change its own sensitivity if it is preyed upon by bats with different echolocating calls. Such is the case of the Australian noctuid moth, Speiredonia spectans, which adapts its acoustic sensitivity according to the characteristics of the call of the bat inside the cave with them.

Ultrasound Avoidance in Crickets

Figure: *A adult male and a juvenile male of the species* Gryllus bimaculatus.

Crickets are preyed on by bats during the night while they fly from one place to another. Avoidance behaviours by crickets were first reported in 1977 by Popov and Shuvalov. They also demonstrated that crickets, like moths, fly away from bats once they've heard their echolocating calls, an example of negative phonotaxis. The cricket will steer itself away from the source of the sound within a very short time frame (40–80 ms) The response is evoked by brief ultrasonic pulses in the 20 to 100 kHz range, pulses which fall within the range of bat ultrasonic echolocating calls (20–100 kHz).

As opposed to moths, the cricket ear, located in the foreleg, is complex- having 70 receptors that are arranged in a tonotopic fashion. This is understandable since crickets don't only need to listen to bats, but also to each other. Crickets have broad frequency sensitivity to different types of echolocating calls. One specific auditory interneuron, the AN2 interneuron, exhibits remarkably rapid responses to echolocating call stimuli. All these receptors synapse on a far lower number of interneurons that relay the receptors' information to the cricket's central nervous system. In the "Teleogryllus" cricket, two ascending interneurons carry information to the brain- one carries information about cricket song (around 5 kHz) while the other gets excited at ultrasound and other high frequencies (15–100 kHz). The ultrasound-sensitive interneuron- labeled INT-1- has been demonstrated as both necessary and sufficient for negative phonotaxis by Nolen and Hoy in 1984: Stimulating int-1 by current injection is sufficient to initiate negative phonotaxis, while hyperpolarising int-1 effectively cancels the turning response to ultrasound. Due to this, int-1 has been proposed to be a command neuron of sorts; in the cricket, int-1 is a bat detector when the cricket is in flight and the interneuron's activity reaches a specific threshold. If these conditions are met, the magnitude of the sound is linearly proportional to the magnitude of the avoidance response. This research also demonstrated that the brain is necessary for the response, since decapitated crickets will fly, but show no avoidance response behaviours.

Bats may have potentially found ways to get around this system. In the Teleogryllus oceanicus cricket, its broad sensitivity can be circumvented by the use of frequency-mismatched calls by part of bats like the gleaning bat, N. Geoffroyi. Furthermore, it has been found that the ultrasound avoidance response seen in crickets during flight is restricted to when the crickets are in flight: that is, the response is extinguished when the crickets are on the ground. It has also been shown that short-winged crickets are less sensitive to ultrasound, but not to low frequencies, than their long-winged counterparts in a wing-dimorphic cricket, "Grillus texensis". A hormone, named 'juvenile hormone' (JH), is believed to play a role in whether the individual develops shorter or longer wings: if the individual has a higher level of JH, its wings will be shorter.

Ultrasound Avoidance in Other Insects

In praying mantises, ultrasound avoidance behaviours are non-directional turns or power dives that are very effective in preventing capture by bats. The mantis ear, located in the midline between the metathoracic (third) legs, comprises two tympana within an auditory chamber that enhances sensitivity. A bilaterally symmetrical pair of auditory interneurons, 501-T3, accurately track the ultrasonic calls during the early stages of a bat attack. Because 501-T3 stops firing just before the evasive response starts, it may be involved in triggering the behaviour. The praying mantis ear first appeared ca. 120 MYA, predating the appearance of echolocating bats by ca. 50 million years, so its original function must be different from its current one.

Arctiid moths use a very different, but highly effective defense against bats. They produce loud ultrasonic clicks in response to ultrasound. Depending on the species of moth and its ecology, the clicks may work by startling the bat, by jamming its echolocation system, or by warning of distastefulness (aposematic).

Green lacewings (Chrysopidae) have sensitive ears on their wings. Ultrasound causes flying lacewings to fold their wings and drop, an effective maneuver for evading capture by bats. Some tettigoniids use a similar strategy, although other species respond much like crickets. Several other insects have sensitive ultrasonic hearing that probably is used in bat evasion, but direct evidence is not yet available. These include scarab beetles, tiger beetles, a parasitoid fly (Ormia sp.).

Orienting

Animal Movement

Predictable fluctuations in environmental conditions impact the behaviour and movement patterns of many animals. When northern temperate ponds freeze over in winter, bald eagles are cutoff from their prey, forced to travel in search of open water. Animals respond by travelling to such seasonal changes in carrying capacity by seeking out abundant resources in the summer and emigrating from limited, harsh resource conditions in the winter. Migration refers to these regular, seasonal journeys of animals that alternate between breeding and non-breeding locations. Many species are able to move great distances and excel at an ability to find their way. A combination of cues and mechanisms are employed to complete some of the most amazing achievements.

Animal Movements and Tinbergen's Four Aims

A series of independent questions can be addressed to explore the causation of migratory movements in animals, such as song birds.

Proximate Causation: hormonal influences on timing (restlessness), cues for navigation.

Ultimate Causation: For functional explanations of songbird migration, migrants must exceed local residents in the number of offspring, despite making up for additional energy costs and dangers associated with migration.

Phylogeny: Many species of birds migrate. Fossil evidence suggests that early songbirds evolved in the tropics. Migration may have arisen when individuals moved north (or southward) in spring to exploit longer daylight hours, local resource abundance in temperate zones, and release from competition in the tropics during the breeding season with its high food demands. The precise course of current major flyways have probably evolved since the last glaciation.

Ontogeny: learning of migration paths and innate preference for migration direction (European warblers, Starlings)

Homing

Figure: *Homing pigeon*

Homing refers to an individual's ability to return from a distant location to the only place it is certain to know something about. Animals may utilise a variety of cues and mechanisms to determine the location of its home site. Limpets fit their shell to one specific spot on a rock. A close fit is critical to withstand the impact of waves when submerged and prevent desiccation when exposed. Scraping algae from the rock surface, they travel from their home site, leaving behind a chemical trail. By reversing direction on this trail they are able to return to the very site they came from.

The homing pigeon has been selectively bred to be able to return to its home over extremely long distances. As a pigeon generally returns to its own nest and its own mate, human handlers have selectively bred subsets of birds that were particularly good at this task. Homing flights from as far as 1000 kilometers have been recorded by exceptional birds at average flying speeds of around 50km/h. For thousands of years humans have used homing pigeons to return messages back to home base by attaching them to the birds legs.

Migration

Migration includes all movements made in response to changes in food availability, habitat or weather.

Nomadic Migration'

'Seasonal Migration': Travel between a summer area for breeding and an area for overwintering are driving the seasonal migration of many animals, such as song birds. The energetic advantages of breeding in a site of seasonally high food resources and extended daylight hours allow diurnal birds to produce larger clutches than related non-migratory species that remain in the tropics year-round.

In autumn, as the days shorten and food sources dwindle, birds return to warmer regions where the available food supply varies little with the season. The success of migrants indicates that these advantages greatly offset the high stress, energetic costs, and other risks of the migration. Cardinals are common winter residents in northern latitudes. As members of a taxon with mostly tropical distribution, they may have reduced their winter return migration.

Mechanisms

Orienting

Many individuals show a distinct preference for departing in one particular direction. This applies to diurnal movements of plankton that feed at the surface during the veening and night, and sink to lower, darker regions to avoid falling prey to visual predators.

Navigation

Navigation refers to the ability of planning and controlling one's movement from one place to another. It requires knowing one's current location on earth, the location of the desired goal, and an ability to calculate travel directions to get from here to there.

Pilotage

Dead Reckoning: Many birds determine a particular flight direction based on environmental cues that serve as landmarks on the Earth's surface, including longitudinally magnetic lines, the lcoation and direction of the sun and other celestial objects, or prevailing wind direction and odors.

In vector navigation individuals arrive at a given destination by maintaining a compass direction (or directions) for a predetermined amount of time or distance. Dead reckoning is the process of estimating one's present position by projecting course and speed from a known past position. Honey bees for instance communicate the location of profitable nectora sources to hive mates. Using a waggle dance they are able to transmit codes for direction and distance.

European Starlings from much of continental Europe overwinter on the channels coast of France and Britain. A group of banded individuals was moved from one part of Europe to a more western geographic location. This transfer would now require a ninety degree change in direction to reach their traditional wintering grounds.

Adult starlings were admirably able to perform that course adjustment and returned at their normal wintering range. Newly-hatched juveniles, however, which had never made the trip before, continued to fly their traditional southwesterly direction and ended up on the coast of Spain.

Desert ants (*Cataglyphis*) are able to obtain an accurate, mean return vector that reverses the distances accumulate in a number of invidual outward movements.

Orientation Cues

Niko Tinbergen's work on Digger wasps illustrates the importance of local visual cues to pinpoint the exact location of the burrow.

Radar images show that migrating bird flocks drift off courses in a strong crosswind, except for flocks that travel along distinct landmarks, such as rivers, coast lines, or mountain ridges. Landmarks may also arise from specific atmospheric conditions and distinct geographic features, such as Point Pelee, which jutts into Lake Erie and serves as key staging ground for migrating hawks in search of updrafts.

Visual Orientation refers to the use of geographic landmarks (rivers, coastlines, ridges, etc.). Diurnal migrants often follow such landmarks, generally of lesser importance for nocturnal migrants, although there are documented cases of nocturnal migrants following rivers or coastlines. This may be particularly important for navigation to a precise breeding or wintering locality.

Cognitive Map

Is an internal representation of the spatial relationships between objects in an animal's surroundings

Carrying Capacity

The carrying capacity of a biological species in an environment is the maximum population size of the species that the environment can sustain indefinitely, given the food, habitat, water and other necessities available in the environment. In population biology, carrying capacity is defined as the environment's maximal load, which is different from the concept of population equilibrium.

For the human population, more complex variables such as sanitation and medical care are sometimes considered as part of the necessary establishment. As population density increases, birth rate often decreases and death rate typically increases. The difference between the birth rate and the death rate is the "natural increase". The carrying capacity could support a positive natural increase, or could require a negative natural increase.

Thus, the carrying capacity is the number of individuals an environment can support without significant negative impacts to the given organism and its environment. Below carrying capacity, populations typically increase, while above, they typically decrease. A factor that keeps population size at equilibrium is known as a regulating factor. Population size decreases above carrying capacity due to a range of factors depending on the species concerned, but can include insufficient space, food supply, or sunlight. The carrying capacity of an environment may vary for different species and may change over time due to a variety of factors, including: food availability, water supply, environmental conditions and living space.

The origins of the term carrying capacity are uncertain with researchers variously stating that it was used "in the context of international shipping" or that it was first used during 19th Century laboratory experiments with micro-organisms. A recent review finds the first use of the term in an 1845 report by the US Secretary of State to the Senate.

Examples

One of the world's best-studied predator-prey relationships is the moose and wolf population of Isle Royale National Park [1] in Lake Superior. Without the wolves, the moose would overgraze the island's vegetation. Without the moose, the wolves would die. The first scientists who studied the issue thought that the wolves would eventually overpopulate and kill all the moose calves, then die from famine. This has not occurred as inbreeding, disease and environmental factors have limited the wolf population naturally.

Easter Island has been cited as an example of a human population crash. When fewer than 100 humans first arrived, the island was covered with trees with a large variety of food types. In 1722, the island was visited by Jacob Roggeveen, who estimated a population of 2000 to 3000 inhabitants with very few trees, "a rich soil, good climate" and "all the county was under cultivation". Half a century later, it was described as "a poor land" and "largely uncultivated".

The ecological collapse which followed has been variously attributed to overpopulation, slave traders, European diseases (including a smallpox epidemic which killed so many so quickly, the dead were left unburied and a tuberculosis epidemic which wiped out a quarter of the population), social upheaval and invasive species (such as the Polynesian rats which may have wiped out the ground nesting birds and eaten the palm tree seeds). Whatever the combination of factors, only 111 inhabitants were left on the island in 1877. For whatever reasons (whether Moai worship, survival, status or sheer ignorance), the question of how many humans the island could realistically support never seems to have been answered.

The Chincoteague Pony Swim is a human-assisted example.

Both herds are managed differently. The National Park Service owns and manages the Maryland herd while the Chincoteague Volunteer Fire Company owns and manages the Virginia herd. The Virginia herd, referred to as the "Chincoteague" ponies, is allowed to graze on Chincoteague National Wildlife Refuge, through a special use permit issued by the U.S. Fish and Wildlife Service. The size of both herds is restricted to approximately 150 adult animals each in order to protect the other natural resources of the wildlife refuge.

A further example is the Island of Tarawa, where the finite amount of space is evident, especially since landfills cannot be dug to dispose of solid waste, due to constraints in the subsurface rock and lack of topographic elevations. With colonial influence and an abundance of food (relative to life before the year 1850), the population has expanded to the extent that overpopulation is transparently present.

Mathematics

The Lotka-Volterra equations are simple mathematical model of population dynamics which show how in a closed system, like that of the wolves and moose on Isle Royale, limited prey will cause the predator population to decline rapidly. An extended example can be used where multiple species are competing for the same resources, or single species feed on multiple prey.

Humans

The application of carrying capacity for the human population has been criticized for not successfully capturing the multi-layered processes between humans and the environment, which have a nature of fluidity and non-equilibrium, and that it often has a blame-the-victim framework. Supporters of the concept argue that humans, like every species, have a finite carrying capacity. Animal population size, living standards, and resource depletion vary, but the concept of carrying capacity still applies. The carrying capacity of Earth has been studied by computer simulation models like World3.

Food Supply and Consumption

Carrying capacity, at its most basic level, is about organisms and food supply, where "X" amount of humans need "Y" amount of food to survive. If the humans neither gain or lose weight in the long run, the calculation is fairly accurate. If the quantity of food is invariably equal to the "Y" amount, carrying capacity has been reached. Humans, with the need to enhance their reproductive success (see Richard Dawkins' *The Selfish Gene*), understand that food supply can vary and also that other factors in the environment can alter humans' need for food. A house, for example, might mean that one does not need to eat as much to stay warm as one otherwise would. Over time, monetary transactions have replaced barter and local production, and consequently modified local human carrying capacity. However, purchases also impact regions thousands of miles away. For example, carbon dioxide from an automobile travels to the upper atmosphere. This led Paul R. Ehrlich to develop the IPAT equation

$$I = P * A * T$$

where:

I is the impact on the environment resulting from consumption

P is the population number

A is the consumption per capita (affluence)

T is the technology factor

Technology is an important factor in the dynamics of carrying capacity. For example, the Neolithic revolution increased the carrying capacity of the world relative to humans through the invention of agriculture. Currently, the use of fossil fuels has artificially increased the carrying capacity of the world by the use of stored sunlight, albeit at many other expenses. Other technological advances that have increased the carrying capacity of the world relative to humans are:

polders, fertilizer, composting, greenhouses, land reclamation, and fish farming. Agricultural capability on Earth expanded in the last quarter of the 20th century. But now there are many projections of a continuation of the decline in world agricultural capability (and hence carrying capacity) which began in the 1990s. Most conspicuously, China's food production is forecast to decline by 37% by the last half of the 21st century, placing a strain on the entire carrying capacity of the world, as China's population could expand to about 1.5 billion people by the year 2050. This reduction in China's agricultural capability (as in other world regions) is largely due to the world water crisis and especially due to mining groundwater beyond sustainable yield, which has been happening in China since the mid-20th century.

Ecological Footprint

One way to estimate human demand compared to ecosystem's carrying capacity is "Ecological Footprint" accounting. Rather than speculating about future possibilities and limitations imposed by carrying capacity constraints, Ecological Footprint accounting provides empirical, non-speculative assessments of the past. It compares historically regeneration rates (biocapacity) against historical human demand (Ecological Footprint) in the same year. One result shows that humanity's demand for 1999 exceeded the planet's biocapacity for 1999 by over 20 percent.

Overpopulation in Wild Animals

Overpopulation in wild animals occurs when a population of a wild species exceeds the carrying capacity of its ecological niche. Overpopulation is a function of the number of individuals compared to the relevant resources, such as the water and essential nutrients they need to survive. In the wilderness, the problem of animal overpopulation is solved by predators. Predators tend to look for signs of weakness in their prey, and therefore usually first eat the old or sick animals. This has the side effects of ensuring a strong stock among the survivors, and controlling the population.

In the absence of predators, animal species are bound by the resources they can find in their environment, but this does not necessarily control overpopulation. In fact, an abundant supply of resources can produce a *population boom* that ends up with more individuals than the environment can support. In this case, starvation, thirst and sometimes violent competition for scarce resources may effect a sharp reduction in population in a very short lapse (a *population crash*). Lemmings, as well as other less popular species of rodents, are known to have such cycles of rapid population growth and subsequent decrease. Some animal species seem to have a measure of self-control, by which individuals refrain from mating when they find themselves in a crowded environment. This voluntary abstinence may be induced by stress or by pheromones.

In an ideal setting, when animal populations grow, so do the number of predators that feed on that particular animal. Animals that have birth defects or weak genes (such as the runt of the litter) also die off, unable to compete over food with stronger, healthier animals.

In reality, an animal that is not native to an environment may have advantages over the native ones, such being unsuitable for the local predators. If left uncontrolled, such an animal can quickly overpopulate and ultimately destroy its environment.

Examples of animal overpopulation caused by introduction of a foreign species abound:

- In the Argentine Patagonia, for example, European species such as the trout and the deer were introduced into the local streams and forests, respectively, and quickly became a plague, competing with and sometimes driving away the local species of fish and ruminants.
- In Australia, when rabbits were introduced by European immigrants, they bred out of control and ate the farm crops and food that both native and farm animals needed. Farmers hunted the rabbits, and also brought cats in to guard against rabbits and rats. These introduced cats created another problem, becoming predators of local species.

Examples of animal overpopulation caused by natural cyclic variations include:

- the 2004 Locust Outbreak in West and North Africa.
- the Australian locust plagues.
- Palestine's eight month long 1915 locust plague.

Overpopulation in Companion Animals

The phenomenon of overpopulation in companion animals refers to the large number of homeless domestic cats and dogs. In the United States alone, between 3 and 4 million cats and dogs are euthanised each year because no one volunteers to adopt them. As a result, most humane societies, animal shelters and rescue groups urge animal caregivers to have their animals spayed or neutered to prevent the births of unwanted and accidental litters.

Effects Upon Animals

Unwanted and stray dogs and cats suffer from neglect and abandonment, deplorable living conditions, insufficient or nonexistent veterinary care, and substandard veterinary practices. Such animals are often victimised by people who treat them inhumanely, due to poverty, lack of knowledge of how to provide care, absence of animal welfare legislation and enforcement, apathy, personal beliefs, and intentional cruelty.

Purebred Preference

One contributing factor in companion animal homelessness is cultural preference for young, purebred animals. Many people prefer purebred, young, healthy animals, and choose to purchase animals of said description, often at significant cost, from breeders. That breeders continue to breed animals while animals are killed in shelters each year is a point of contention for many animal rights groups.

Recognising the high demand for purebred animals, some people choose to engage in backyard breeding or operate puppy mills, practices where people breed purebred animals for profit, often without concern for the health or welfare of any of the animals involved. These animals may be sold through pet stores or directly from the breeders themselves.

Additionally, individuals seeking purebred animals may not realize that a homeless animal adopted from a shelter can have many advantages: often the shelter will have performed all necessary veterinary procedures, such as spaying or neutering, vaccination, deworming, microchipping, etc. Also, the personality of a kitten or puppy is not always an indicator of how the animal will behave in adulthood. Many shelter animals have reached adulthood and their personalities are apparent, allowing the would-be caregiver to select an animal with a personality that suits them.

Global Effects

Dealing with a population of unwanted companion animals is a major concern to animal welfare and animal rights groups. Companion animal overpopulation can also be an ecological concern. It is also a financial problem: capturing, impounding and eventual euthanasia costs taxpayers and private agencies millions of dollars each year.

Omnivore

Omnivores (from Latin: *omni* all, everything; *vorare* to devour) are species that eat both plants and animals as their primary food source. They are opportunistic, general feeders not specifically adapted to eat and digest either meat or plant material primarily.

Figure: *Ravens are omnivores.*

Definition

Although the term *omnivore* literally means *eater of everything*, omnivores cannot really eat "everything" that other animals eat; they can only eat things that are at least moderately easy to get and still at least moderately nutritious. For example, most of them cannot live by grazing (easy to get, but not nutritious enough), nor can they eat some hard-shelled animals or successfully hunt large or fast prey (nutritious, but too hard to get).

Evolutionary History

Omnivory has evolved several times among different groups of animals. The first vertebrates were piscivores, then insectivores, carnivores and finally herbivores. A complex set of adaptations was necessary for feeding on highly fibrous plant materials, requiring structural modifications to the teeth, jaws, and digestive tract. Only a small proportion of extant tetrapods are obligate herbivores; many carnivores also consume low-fiber plant material as well as insects and fish, so it could be that early tetrapods made the transition to fully fledged herbivory by way of omnivory. There were no omnivores at the time of ice age.

Omnivorous Species

Although there are cases of Carnivores eating plant matter, as well as examples of herbivores eating meat, the classification refers to the adaptations and main food source of the species in general, so these exceptions do not make either individual animals nor the species as a whole omnivores. Pigs furnish a well-known example of an omnivore. Crows are another example of an omnivore that many people see every day. As a species, humans are currently divided. While they do not share many of the characteristics of most other omnivores, such as claws or fangs or short digestive tracts, the majority of humans eat as omnivores, eating both plants and animals for food on a daily basis.

Figure: *Most bears are omnivores, with the exception of polar bears and pandas.*

Most bear species are considered omnivores, but individual diets can range from almost exclusively herbivorous to almost exclusively carnivorous, depending on what food sources are available locally and seasonally. Polar bears are classified as carnivores while pandas are classified as herbivores, although giant pandas will eat some meat (e.g., insects) from time to time, and polar bears will sometimes eat plants (such as kelp) but of neither is the exception a significant part of their diet.

Various mammals are omnivorous by nature, such as pigs, badgers, bears, coatis, hedgehogs, opossums, skunks, sloths, squirrels, raccoons, chipmunks, mice and rats. Also some primates are omnivorous including chimpanzees and baboons. Various birds are omnivorous, whose diet varies from berries and nectar to insects, worms, fish, and small rodents; examples include cassowarys, chickens, crows and related corvids, keas, rallidae, and rheas. In addition, some lizards, turtles, fish, such as piranhas, and invertebrates are also omnivorous.

While virtually all mammals may display "omnivorous" behaviour patterns—depending on conditions of supply, culture, etc.—mammals will generally prefer one class of food or another, with optimised digestive processes. Like most arboreal species, most squirrels are primarily granivores, preferring nuts and seeds. but as with virtually all mammals, squirrels can resort to consuming some meat as fallback food if starving or facultatively, e.g., when nests are in danger of being raided by predators, etc. Depending on the species of bear, there is generally a preference for one class of food or another as plants and animals are digested differently. While scientific classification aims to promote communication and analysis of various differences and similarities between species, the concept of an "omnivore" is broad and could be applied to virtually any mammal since disease risks and the quality of digestion are often not considered. There are social, psychological and non-nutritive factors that influence diet behaviour. "[T]he behavioural basis of omnivory has not been thoroughly explored... and food selection behaviour is central to understanding the causes and consequences of omnivory. However, few studies have actually addressed this issue through rigorous tests of multiple hypotheses."

In order for the concept of "omnivore" to be regarded as a scientific classification, some clear set of measurable and relevant criteria would need to be considered to differentiate between an "omnivore" and the other vague but less ambiguous diet categories e.g., faunivore, folivore, scavenger, etc.

Mesopredator Release Hypothesis

The mesopredator release hypothesis is a relatively new hypothesis from 1988 which describes the phenomenon of trophic cascade in certain terrestrial communities. It states that as top predators decline in an ecosystem, an increase in the populations of mesopredators occurs. Mesopredators are middle trophic level predators such as raccoons, skunks, snakes, cownose rays, and small sharks.

Hypothesis

Mesopredator release is a fairly new concept that is gaining approval, although it is still being debated and studied. The mesopredator release hypothesis states that if an apex predator is taken out of an ecosystem, the number of mesopredators (defined as medium-sized predators, such as raccoons, skunks, snakes, cats, and foxes) often increase in abundance when larger predators are eliminated. Mesopredator populations will surge and the predation of smaller, more vulnerable prey species will increase. As a result, the shared prey may suffer more from predation than when the apex predators (or top-predators) were controlling the mesopredators.

This may lead to dramatic prey population decline, or even extinction, especially on islands. This process arises when mammalian top predators are considered to be the most influential factor on trophic structure and biodiversity in terrestrial ecosystems. Top predators may feed on herbivores and kill predators in lower trophic levels as well. Thus, reduction in the abundance of top predators may cause the medium-sized predator population to increase, therefore having a negative effect on the underlying prey community.

The mesopredator release hypothesis offers an explanation for the abnormally high numbers of mesopredators and the decline in prey abundance and diversity. The hypothesis supports the argument for conservation of top predators because they protect smaller prey species that are in danger of extinction. This argument has been a subject of interest within conservation biology for years, but few studies have proved the phenomenon.

Criticism

One of the main criticisms of the mesopredator release hypothesis is that it argues in favourof the top-down control concept and excludes the possible impacts that bottom-up control could have on higher trophic levels. This means that it supports the argument that top predators control the structure and population dynamics of an ecosystem, but it does not take into account that prey species and primary producers also have an effect on the ecosystem's structure.

Another major problem is that the hypothesis is offered as an explanation after large predators have already become rare or extinct in an ecosystem. Consequently, there is no data on the past ecosystem structure and the hypothesis cannot be tested. As a result, information on the past conditions has been inferred from studies of the present conditions. However, contemporary examples of mesopredator release exist, such as the culling of cats on Macquarie Island.

The hypothesis is sometimes also applied to humans as apex predators that produce top-down effects on lower trophic levels. However, it fails to recognise bottom-up effects that anthropogenic land transformations can have on landscapes on which primary producers, prey species, and mesopredators dwell. Possible bottom-up effects on an ecosystem can be from bioclimatic impacts on ecosystem productivity and from anthropogenic habitat alterations.

Examples of anthropogenic habitat change include agriculture, grazing land, and urbanisation. More importantly, the hypothesis does not take into account that higher trophic levels are affected by primary productivity. It also does not mention that trophic interactions operate at different strengths according to the ecosystem.

Therefore, the roles of predation and food/nutrient processes in influencing ecosystem structures remain open to controversy and further testing.

Pair Bond

Figure: *Northern Gannet pair*

In biology, a pair bond is the strong affinity that develops in some species between the males and females in a pair, potentially leading to breeding. Pair-bonding is a term coined in the 1940s that is frequently used in sociobiology and evolutionary psychology circles. The term often implies either a lifelong socially monogamous relationship or a stage of mating interaction in socially monogamous species. It is sometimes used in reference to human relationships.

Monogamous voles, such as prairie voles, have significant differences in the density and distribution of vasopressin receptors in their brain when compared to polygamous voles. These differences are located in the ventral forebrain and the dopamine-mediated reward pathway. Both vasopressin and dopamine act in this region to coordinate rewarding activities such as mating, and regulate selective affiliation. These species specific differences have shown to correlate with social behaviours, and in monogamous prairie voles are important for facilitation of pair bonding.

Varieties

According to evolutionary psychologists David Barash and Judith Lipton, from their 2001 book *The Myth of Monogamy*, there are several varieties of pair bonds:

- *Short-term pair-bond:* a transient mating or associations
- *Long-term pair-bond:* bonded for a significant portion of the life cycle of that pair
- *Lifelong pair-bond:* mated for life
- *Social pair-bond:* attachments for territorial or social reasons, as in cuckold situations

- *Clandestine pair-bond:* quick extra-pair copulations
- *Dynamic pair-bond*: e.g. gibbon mating systems being analogous to "swingers".

Examples

When discussing the social life of the bank swallow, Lipton and Barash state:

For about four days immediately prior to egg-laying, when copulations lead to fertilizations, the male bank swallow is very busy, attentively guarding his female. Before this time, as well as after—that is, when her eggs are not ripe, and again after his genes are safely tucked away inside the shells—he goes seeking extra-pair copulations with the mates of other males...who, of course, are busy with defensive mate-guarding of their own.

In various species, males provide parental care and females mate with multiple males. For example, recent studies show that extra-pair copulation frequently occurs in monogamous birds in which a "social" father provides intensive care for its "social" offspring. A University of Florida scientist reports that male sand gobies work harder at building nests and taking care of eggs when females are present – the first time such "courtship parental care" has been documented in any species.

Antomy of Dog

Dog anatomy includes the same internal structures that are in humans. Details of structures vary tremendously from breed to breed, more than in any other animal species, wild or domesticated, as dogs vary from the tiny Chihuahua to the giant Irish Wolfhound.

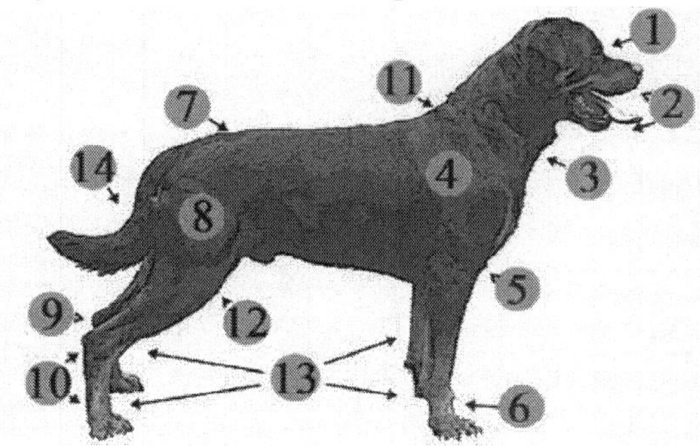

Figure: *External anatomy (topography) of a typical dog: 1. Stop 2. Muzzle 3. Dewlap (throat, neck skin) 4. Shoulder 5. Elbow 6. Forefeet 7. Croup (rump) 8. Leg (thigh and hip) 9. Hock 10. Hind feet 11. Withers 12. Stifle 13. Paws 14. Tail*

Physical Characteristics

Like most predatory mammals, the dog has powerful muscles, a cardiovascular system that supports both sprinting and endurance, and teeth for catching, holding, and tearing.

The dog's ancestral skeleton provided the ability to jump and leap. Their legs are designed to propel them forward rapidly, leaping as necessary, to chase and overcome prey. Consequently, they have small, tight feet, walking on their toes; their rear legs are fairly rigid and sturdy; the front legs are loose and flexible, with only muscle attaching them to the torso.

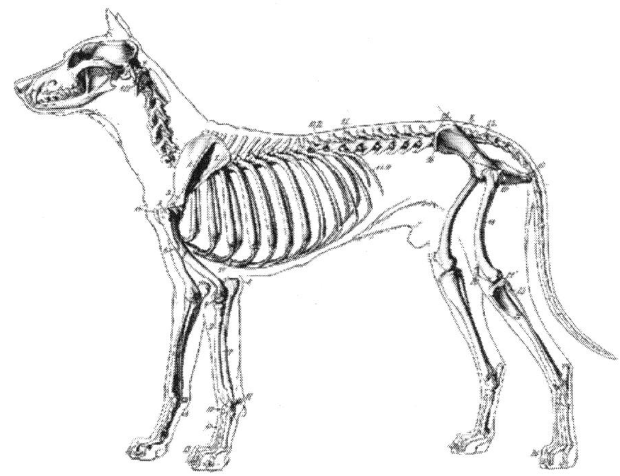

Figure: *Skeleton of a domestic dog*

Although selective breeding has changed the appearance of many breeds, all dogs retain the basic ingredients from their distant ancestors. Dogs have disconnected shoulder bones (lacking the collar bone of the human skeleton) that allow a greater stride length for running and leaping. They walk on four toes, front and back, and have vestigial dewclaws on their front legs and sometimes on their rear legs.When a dog has extra dewclaws in addition to the usual one on each front leg, the dog is said to be "double dewclawed".

There is some debate about whether a dewclaw helps dogs to gain traction when they run because, in some dogs, the dewclaw makes contact when they are running and the nail on the dewclaw often wears down in the same way that the nails on their other toes do, from contact with the ground.

However, in many dogs the dewclaws never make contact with the ground; in this case, the dewclaw's nail never wears away, and it is then often trimmed to keep it to a safe length.

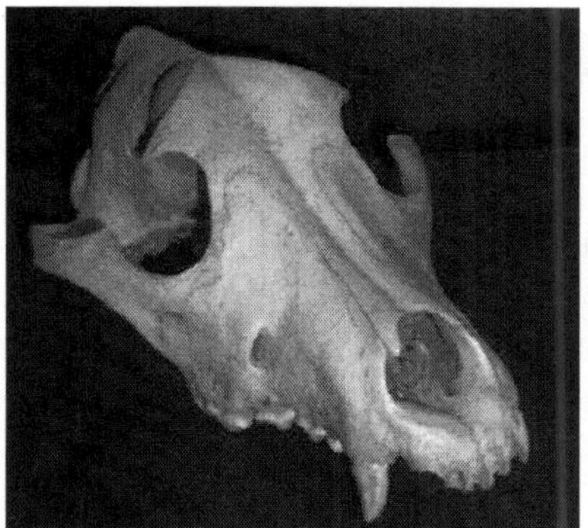

Figure: *Skull of a dog*

The dewclaws are not dead appendages. They can be used to lightly grip bones and other items that dogs hold with the paws. However, in some dogs these claws may not appear to be connected to the leg at all except by a flap of skin; in such dogs the claws do not have a use for gripping as the claw can easily fold or turn.

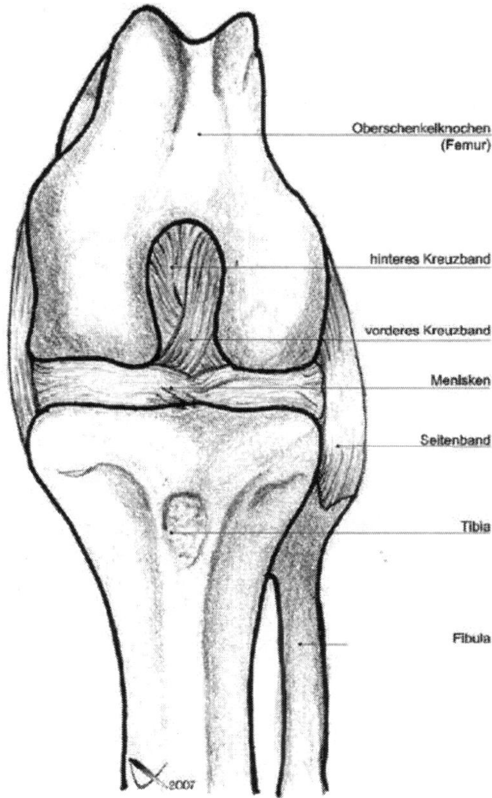

Figure: *Dog knee*

There is also some debate as to whether dewclaws should be surgically removed.The argument for removal states that dewclaws are a weak digit, barely attached to the leg, so that they can rip partway off or easily catch on something and break, which can be extremely painful and prone to infection. Others say the pain of removing a dewclaw is far greater than any other risk. For this reason, removal of dewclaws is illegal in many countries. There is, perhaps, an exception for hunting dogs, who can sometimes tear the dewclaw while running in overgrown vegetation. If a dewclaw is to be removed, this should be done when the dog is a puppy, sometimes as young as 3 days old, though it can also be performed on older dogs if necessary (though the surgery may be more difficult then). The surgery is fairly straight-forward and may even be done with only local anesthetics if the digit is not well connected to the leg. Unfortunately many dogs can't resist licking at their sore paws following the surgery, so owners need to remain vigilant.

In addition, for those dogs whose dewclaws make contact with the ground when they run, it is possible that removing them could be a disadvantage for a dog's speed in running and changing of direction, particularly in performance dog sports such as dog agility.

The dog's ancestor was about the size of a Dingo, and its skeleton took about 10 months to mature. Today's toy breeds have skeletons that mature in only a few months, while giant breeds such as the Mastiffs take 16 to 18 months for the skeleton to mature. Dwarfism has affected the proportions of some breeds' skeletons, as in the Basset Hound.

Knowledge of basic anatomy also helps when competing in dog shows or contests.

Size

Researchers have identified a particular piece of genetic material that is common to every small-dog breed and, in turn, is probably responsible for making them tiny. The study, published in 2007, found a regulatory sequence (not a gene) next to the gene IGF1; together the gene and regulatory sequence together are known as a haplotype that "is a major contributor to body size in all small dogs." Medium and large size dogs do not usually have the regulatory sequence, although the small-size sequence was found in the Rottweiler breed. The study included 3,241 dogs from 143 breeds. The researchers concluded the genetic instructions to make dogs small must be at least 12,000 years old, and it is not found in wolves. Another study has shown that lap dogs (small dogs) are among the oldest dog types.

Modern dog breeds show more variation in size, appearance, and behaviour than any other domestic animal. Within the range of extremes, dogs generally share attributes with their wild ancestors, the wolves. Dogs are predators and scavengers, possessing sharp teeth and strong jaws for attacking, holding, and tearing their food. Although selective breeding has changed the appearance of many breeds, all dogs retain basic traits from their distant ancestors. Like many other predatory mammals, the dog has powerful muscles, fused wristbones, a cardiovascular system that supports both sprinting and endurance, and teeth for catching and tearing.

Sight: Like most mammals, dogs are dichromats and have colour vision equivalent to red-green colour blindness in humans. Different breeds of dogs have different eye shapes and dimensions, and they also have different retina configurations. Dogs with long noses have a "*visual streak*" which runs across the width of the retina and gives them a very wide field of excellent vision, while those with short noses have an "*area centralis*" — a central patch with up to three times the density of nerve endings as the *visual streak* — giving them detailed sight much more like a human's.

Some breeds, particularly the sighthounds, have a field of vision up to 270° (compared to 180° for humans), although broad-headed breeds with short noses have a much narrower field of vision, as low as 180°.

Hearing: Dogs detect sounds as low as the 16 to 20 Hz frequency range (compared to 20 to 70 Hz for humans) and above 45 kHz (compared to 13 to 20 kHz for humans),and in addition have a degree of ear mobility that helps them to rapidly pinpoint the exact location of a sound. Eighteen or more muscles can tilt, rotate and raise or lower a dog's ear. Additionally, a dog can identify a sound's location much faster than a human can, as well as hear sounds up to four times the distance that humans are able to. Those with more natural ear shapes, like those of wild canids like the fox, generally hear better than those with the floppier ears of many domesticated species.

Smell: Dogs have nearly 220 million smell-sensitive cells over an area about the size of a pocket handkerchief (compared to 5 million over an area the size of a postage stamp for humans). According to nhm.org, dogs can sense odours at concentrations nearly 100 million times lower than humans can. According to Dummies.com, the percentage of the dog's brain that is devoted to analysing smells is actually 40 times larger than that of a human. Some dog breeds have been selectively bred for excellence in detecting scents, even compared to their canine brethren.

Coat (Dog)

The coat of the domestic dog (Canis lupus familiaris) refers to the hair that covers its body. A dog's coat may be a *double coat*, made up of a soft undercoat and a coarser topcoat, or a *single coat*, which lacks an undercoat. The terms fur and hair are often used interchangeably when describing a dog's coat, however in general, a double coat, *e.g.*, like that of the Newfoundland and most mountain dogs, is referred to as a *fur coat*, while a single coat, like that of the Poodle, is referred to as a *hair coat*.

Colours, Patterns, Lengths & Textures

There is a greater variety of coat colours, patterns, lengths and textures found in the domestic dog than in its wolf relations, even though dogs and wolves belong to the same species (Canis lupus). Coat colours in dogs were not likely initially selected for by humans but were probably the inadvertent outcome of some other selection process (i.e. selection for tameness). Research has found that tameness brings associated physical changes, including coat colouring and patterning.

Domestic dogs often display the remnants of countershading, a common natural camouflage pattern. The basic principle of countershading is when the animal is lit from above, shadows will be cast on the ventral side of the body. These shadows could provide a predator or prey with visual cues relating to the movement of the animal. By being lighter coloured on the ventral side of the body, an animal can counteract this, and thereby fool the predator or prey. An alternative explanation is that the dorsal and ventral sides of an animal experience different selection pressures (from the need to blend in to different backgrounds when viewed from above and below) resulting in differing colouration.

Genetic Basis of Colour and Pattern

Modern breeds of dog exhibit a diverse range of coat colourings, patterns, lengths and textures. In recent years, the understanding of the genetic basis for coat colouring and patterning and coat length and texturing has increased significantly.

There are currently eight known genes within the canine genome that are associated with coat colour. Each of these genes occurs in at least two variants, or alleles, which accounts for the variation in coat colour between animals. Each of these genes exists at a fixed location, or locus, of the animal's genome. The loci associated with canine coat colour are:

A (Agouti) Locus

The alleles at the A locus are related to the production of agouti signalling protein (ASIP) and determine whether an animal expresses an agouti appearance, and if so what type, by controlling the distribution of pigment in individual hairs. There are five suspected alleles that occur at the A locus:

- aw= Wild-type agouti (cream to red hair with black tips)
- A^y = Fawn (cream to red hair with darker tips) or sable (solid black hairs interspersed amongst lighter reddish hairs)
- A^s = Saddling
- a^t = Tan points; Tricolours
- a = Recessive black (inhibition of phaeomelanin).

Most texts suggest that the dominance hierarchy for the A locus alleles appears to be as follows: $A^y > A^s > a^w > a^t > a$, however research suggests the existence of pairwise dominance/recessive relationships in different families and not the existence of a single hierarchy in one family. This means, for example, that Asss may be incompletely dominant over a^t.

B (Brown) Locus

The alleles at the B locus are related to the production of tyrosinase related protein 1 (TYRP1) and determine the degree to which an animal expresses tyrosinase, an enzyme related to the production of melanin, in its coat and skin (including the nose and paw pads). There are two known alleles that can occur at the B locus:

- B = Black
- b = Brown (includes several alleles - b^s, b^d and b^c).

B is dominant to b. An animal that has at least one copy of the B allele will have a black nose, paw pads and eye rims while an animal that is homozygous for any of the b alleles will have a liver nose, paw pads and eye rims.

D (Dilute) Locus

The alleles at the D locus (the melanophilin gene or MLPH) are related to the dilution of eumelanin and/or phaeomelanin and determine the intensity of pigmentation. There are two known alleles:

- D = Not Diluted
- d = Diluted (Black becomes grey or blue; brown becomes light tan or "Isabella").

D is dominant to d. Homozygosity of d is sometimes accompanied by hair loss and recurrent skin inflammation, a condition referred to as either colour dilution alopecia (CDA) or black hair follicular dysplasia (BHFD) depending upon the breed of dog.

E (Extension) Locus

The alleles at the E locus (the melanocortin receptor 1 gene or MC1R) determines whether an animal expresses a melanistic mask or a grizzle overlay, as well as determining whether an animal expresses eumelanin in its coat. Expression of eumelanin will result in a black or brown coat, while a lack of expression of eumelanin will result in a red or yellow coat. There are four known alleles that occur at the E locus:

- E = No mask, animal expresses eumelanin (coat will be black or brown)
- E^G = Grizzle (dark overlay covering the top and sides of the body, head and tail, and the outside of the limbs)
- E^m = Mask, animal expresses eumelanin (coat will be black or brown)
- e = No mask, animal does not express eumelanin (coat will be red or yellow)

The dominance hierarchy for the E locus alleles appears to be as follows: $E^m > E^G > E > e$. The Grizzle allele is specific to Salukis and Afghan Hounds, the latter in which it is referred to as "Domino". The expression of E^G is dependant upon the animal being homozygous for a^t and not possessing E^m or K^B. An animal that is homozygous for e will express a red or yellow coat regardless of the alleles at other loci (unless the animal is homozygous for c^a at the C locus in which case it will be albino).

H (Harlequin) Locus

DNA studies have not yet isolcated the gene at the H locus, but the traits associated with it have been mapped to chromosome 9. The H locus is a modifier locus (of the M locus) and the alleles at the H locus will determine if an animal expresses a harlequin pattern (white base with black patches). There are two alleles that can occur at the H locus:

- H = Harlequin
- h = Non-harlequin.

H is dominant to h. Breeding data suggests that H is embryonic recessive lethal and that therefore all harlequins are H/h. The Harlequin allele is specific to Great Danes. As H is a modifier locus of the M locus, in order for the Harlequin pattern to be expressed, one copy of the H allele (at the H locus) and one copy of the M allele (at the M locus) must be present (i.e. H/h and M/m).

K (Dominant Black) Locus

The alleles at the K locus (the â-Defensin 103 gene or DEFB103) determine the colouring pattern of an animal's coat. There are three known alleles that occur at the K locus:

- K^B = Solid colouring (does not mean than white markings can not appear)
- k^{br} = Brindle
- k^y = Enables the expression of agouti alleles that require the expression of phaeomelanin.

The dominance hierarchy for the K locus alleles appears to be as follows: $K^B > k^{br} > k^y$. The colouring of an animal that possesses at least one K^B will be determined by the alleles it possesses at the B and E loci. An animal with one k^{br} allele and no K^B allele will express a brindle pattern to its coat. An animal that is homozygous for k^y will express the agouti pattern in accordance with the alleles it has at the A locus.

M (Merle) Locus

The alleles at the M locus (the SILV gene) determine whether an animal expresses a merle pattern to its coat (patches of sporadic coloured and white hairs and other patches of solid colour). There are two alleles that can occur at the M locus:

- M = Merle (visible in dogs that are not e/e)
- m = Non-merle

M is dominant to m. Both heterozygosity and homozygosity of the merle gene (i.e. M/m and M/M) are linked to a range of auditory and ophthalmologic abnormalities.

S (Spotting) Locus

The alleles at the S locus (the microphthalmia-associated transcription factor gene or MITF) determine the degree and distribution of spotting of an animal's coat. There is disagreement as to the number of alleles that occur at the S locus, with researchers postulating either two or four alleles. The four alleles postulated are:

- S = Solid colour (small areas of white may appear on chest, toes or tail tip)

- s^i = Irish-spotting (white on muzzle, forehead, feet, legs, chest and tail)
- s^p = Pie-bald spotting (large areas of white)
- s^w = Extreme pie-bald spotting (Extremely large areas of white, almost completely white).

S is dominant to s. DNA studies have not yet confirmed the existence of all four alleles, with some research suggesting the existence of at least two alleles (S and s^p) and other research suggesting the possible existence of a third allele (s^i). It has been suggested that what appears to be the result of an s^w allele is in fact the result of plus and minus modifiers acting on one of the other alleles. It is thought that the spotting that occurs in Dalmatians is the result of the interaction of three loci (the S locus, the T locus and F locus) giving them a unique spotting pattern not found in any other breed.

Postulated Colour & Pattern Loci

There are at least five additional theoretical loci thought to be associated with coat colour in dogs. DNA studies are yet to confirm the existence of these genes or alleles but their existence is theorised based on breeding data:

C (Coloured) Locus

The alleles at the theoretical C locus are thought to determine the degree to which an animal expresses phaeomelanin, a red-brown protein related to the production of melanin, in its coat and skin. Five alleles are theorised to occur at the C locus:

- C = Full colour (animal expresses phaeomelanin)
- c^{ch} = Chinchilla (partial inhibition of phaeomelanin resulting in decreased red pigment)
- c^e = Extreme dilution (inhibition of phaeomelanin resulting in extremely reduced red pigment)
- c^b/c^p = Blue-eyed albino/Platinum (almost total inhibition of phaeomelanin resulting in near albino appearance)
- c^a = Albino (complete inhibition of phaeomelanin production, resulting in complete inhibition of melanin production).

The C locus in dogs is not well understood and the theorised alleles are based on those present in other species. True albinism has not been conclusively shown to exist in dogs. It is thought that an animal that is heterozygous for the C allele with one of the other alleles will express a result somewhere between the two alleles.

F (Flecking) Locus

The alleles at the theoretical F locus are thought to determine whether an animal displays small, isolated regions of white in otherwise pigmented regions (not apparent on white animals). Two alleles are theorised to occur at the F locus:

- F = Flecked
- f = Not flecked.

It is thought that F is dominant to f.

G (Progressive Greying) Locus

The alleles at the theoretical G locus are thought to determine if premature greying of the animal's coat will occur. Two alleles are theorised to occur at the G locus:

- G = Premature greying
- g = No premature greying.

It is thought that G is dominant to g.

I (Intensity) Locus

The alleles at the theoretical I locus are thought to affect phaeomelanin expression. Two alleles are theorised to occur at the I locus:

- I = Intense red, not diluted
- i = Not intese red.

It is thought that I and i are co-dominant, so that animals with i/i will be paler than animals with I/i.

T (Ticking) Locus

The alleles at the theoretical T locus are thought to determine whether an animal displays small, isolated regions of pigment in otherwise white regions (not apparent on non-white animals). Two alleles are theorised to occur at the T locus:

- T = Ticked
- t = Not ticked.

It is thought that T is dominant to t.

Genetic Basis of Length & Texture

Research indicates that the majority of variation in coat growth pattern, length and curl can be attributed to mutations in three genes, the R-spondin-2 gene or RSPO2, the fibroblast growth factor-5 gene or FGF5, and the keratin-71 gene or KRT71.

The L (Length) Locus

The alleles at the L locus (the fibroblast growth factor-5 gene or FGF5) determine the length of the animal's coat. There are two known alleles that occur at the L locus:

- L = Short coat
- l = Long coat.

L is dominant to l.

The W (Wired) Locus

The alleles at the W locus (the R-spondin-2 gene or RSPO2) determine the coarseness and the presence of "facial furnishings" (e.g. beard, moustache, eyebrows). There are two known alleles that occur at the W locus:

- W = Wire (hair is coarse and facial furnishings present)
- w = Non-wire (hair is not coarse and facial furnishings are not present).

W is dominant to w. Animals that are homozygous for l (i.e. l/l) and possess at least one copy of W will have long, soft coats with furnishings, rather than wirey coats.

The R (Curl) Locus

The alleles at the R locus (the keratin-71 gene or KRT71) determine whether an animal's coat is straight or curly. There are two known alleles that occur at the R locus:

- R = Straight
- r = Curly.

R is dominant to r.

Interaction of Length & Texture Genes

These three genes responsible for the length and texture of an animal's coat interact to produce seven different phenotypes:

- Short (e.g. Basset Hound)
- Wire (e.g. Australian Terrier)
- Curly-wire (e.g. Airedale Terrier)
- Long (e.g. Golden Retriever)
- Long with furnishings (e.g. Bearded Collie)
- Long and curly (e.g. Irish Water Spaniel)
- Long and curly with furnishings (e.g. Bichon Frisé).

Hair Growth

The coat of most dogs grows to a specific length and then stops growing, while the coats of some dogs grow continuously in a manner similar to human hair growth. Examples of breeds of dog whose coats grow continuously are:

- Bedlington Terrier
- Bichon Frisé
- Irish Water Spaniel
- Kerry Blue Terrier
- Maltese
- Poodle
- Portuguese Water Dog
- Schnauzer
- Soft Coated Wheaten Terrier.

Corded Coats

Corded coats, like those of the Puli and Komondor are thought to be the result of continuously growing curly coats. Other breeds with continuously growing curly coats, such as the Poodle, can also be groomed to cord.

Hairless

Some breeds of dog do not grow hair on parts of their bodies and may be referred to as "hairless". Examples of "hairless" dogs are the Xoloitzcuintli (Mexican Hairless Dog), the Peruvian Inca Orchid (Peruvian Hairless Dog) and the Chinese Crested. Research suggests that hairlessness is caused by one or more dominant alleles, one or more of which is homozygous lethal.

Genetic Testing & Phenotype Prediction

In recent years genetic testing for the alleles of some genes has become available Software is also available to assist breeders in determining the likely outcome of matings.

Tail

There are many different shapes for dog tails: straight, straight up, sickle, curled, cork-screw.

In some breeds, the tail is traditionally docked to avoid injuries (especially for hunting dogs). It can happen that some puppies are born with a short tail or no tail in some breeds.

Puppy Characteristics

Puppies often have characteristics that do not last beyond early puppyhood. Eye colour often changes from blue to its adult colour as the puppy matures. The coat colour may change: Kerry Blue Terrier puppies have black coats at birth and change to blue with maturity, and Dalmatians are white and gain their spots with age. The ear shape will also often change, especially with erect-eared breeds such as the German Shepherd Dog which have soft ears at birth, but the cartilage strengthens with age. Labrador Retrievers and other swimming dogs start off with a very fluffy puppy coat, and over time the water proof layer grows. Puppies that are going to grow into larger dogs will often have over sized paws to begin with, and then the rest of them grow to fit.

Temperature Regulation

It is a common misconception that dogs do not sweat. They do sweat, mainly through the footpads, but only a small fraction of a dog's excess heat is lost this way. Primarily, dogs regulate their body temperature through panting. Panting moves cooling air over the moist surfaces of the tongue and lungs, rejecting heat to the atmosphere.

Dogs possess a rete mirabile, a complex of intermingled small arteries and veins, in the carotid sinus at the base of their neck. This acts to thermally isolate the head, which contains the brain, the most temperature-sensitive organ, from the body, which contains the muscles, where most of the heat is generated. The result is that dogs can sustain intense physical exertion over a prolonged time in a hot environment, compared to animals which lack this apparatus; thus, a dog chasing a jackrabbit through the desert may not be able to outrun the rabbit, but it can continue the chase until the rabbit slows due to overheating.

Behaviour of Dog

Bark (Utterance)

A bark is a noise most commonly produced by dogs and puppies. Other animals that make this noise include wolves and quolls. *Woof* is the most common representation in the English language for this sound (especially for large dogs). Other transliterations include the onomatopoeic *ruff*, *arf*, *au au*, *yip* (for small dogs), and *bow-wow*.

Why Dogs Bark

Although dogs are descended from the wolf, *Canis lupus*, their barking constitutes a significant difference from their parent species.

Although wolves do bark, they do so only in specific situations. According to Coppinger and Feinstein, dogs bark in long, rhythmic stanzas but adult wolf barks tend to be brief and isolated. Compared with wolves, dogs bark frequently and in many different situations.

It has been suggested that the reason for the difference lies in the dog's domestication by humans. An increased tendency to bark could have been useful to humans to provide an early warning system. Domestication has altered the physical appearance of dogs. Dogs present a striking example of neoteny, the retention of juvenile characteristics in the adults. They are similar to young wolves in many of their mannerisms and physical features, such as large heads, flat faces, large eyes, submissiveness and vocalizing – all of which are exhibited in wolf puppies.

Individual dogs bark for a variety of reasons. They may bark to attract attention, to communicate a message, or to express excitement. Statistical analysis has revealed that barks can be divided into different subtypes based on context and that individual dogs can be identified by their barks. Disturbance barks tend to be harsh, low frequency, and unmodulated, whereas isolation and play barks tend to be tonal, higher frequency, and modulated. Barks are often accompanied by body movements as part of a broader package of dog communication.

Types of Barking

Warning Bark: A warning will usually start out as a low, quiet, but ferociously noticeable growl before escalating into something of a howling bark. This type of reaction is most typically seen in domesticated animals in response to a perceived territorial intrusion. The dog may also bare its teeth if it feels immediately threatened.

Also most monkeys make this type of barking noise when communicating with other monkeys or to scare off larger animals.

Alarm Barking

It is a dog's attempt to be alert, attentive, and informative to its human "pack", as regarding unusual events. This kind of barking is known as 'alarm barking', and is common within a variety of breeds. It does not signify aggression, and (although often associated with unusual noises intruding on the dog's 'territory') is not the same as territoriality type barking. It may take the form of just one or a few barks, or it may give rise to sustained barking until the dog sees that some action has been taken. Alarm barking is more likely to arise when a dog can hear, but not see the source of, some noise. Examples of sounds which commonly cause alarm barking include doorbells, cars, noises from adjacent dwellings, and the like. It is a behaviour that tends to develop with age and maturity, and also can be related to whether there are others around who might need to be informed of such events - often an alarm barker will remain quiet if alone and there is nobody to 'tell'.

Barking on Command

A dog can be trained to bark on command.

Playful Bark

Some dogs will bark for a treat or when they are playing. Dogs will also bark when they like what they're playing.

Need Bark

Many dogs can self-develop actions that serve themselves if they need something. This is commonly done along with a form of action. Example, a dog may nudge its bowl and bark if it is hungry, or it may nudge its owner and bark if it wants some form of attention such as petting.

Barking as Nuisance

Bark Control: Nuisance-barking dogs sound off for no particular reason. "Many dogs bark when they hear other dogs barking," said Katherine A. Houpt, V.M.D., Ph.D., director of the Cornell Animal behaviour Clinic. Nuisance, inappropriate, or excessive barking comprises between 13 and 35 percent of behaviour-problem complaints by dog owners, Houpt noted. The electric collars deliver an irritating shock of adjustable intensity when a vibration sensor in the collar detects barking. The citronella collar releases a spray of the plant-based fragrance when a microphone in the collar senses barking. For the eight dogs that wore both types of collars (one shepherd mix did not complete the study), all owners found the citronella collar to be effective in reducing or stopping nuisance barking and most preferred the fragrance spray. Four out of eight owners said electric shocks had no effect on their dogs—they kept on barking. The citronella collars were not without problems, Juarbe-D'az noted. One dog owner complained that citronella oil stained the upholstery when the couch-potato pooch barked. "One owner thought the scent was preferable to her dog's body odor."

Canine barking can be a nuisance to neighbours, and is a common problem dog owners or their neighbours may face. (Many dogs can bark at 100 dBA. Even at 17.5 yards away and with the dog outside a closed window, the noise level of a barking dog can be well over the level that causes psychological distress.) Different kinds of barking often require different kinds of approach to reduction.

Common approaches are as follows:

1. Attempting to understand, and if possible eliminate, the causes of barking.
2. Using positive training methods to correct the behaviour. Dogs may bark from anxiety or stress, so punishment can often cause problems by reinforcing a cycle of bad behaviour. Positive approaches can include:
   - *Repeated exposure* to stimuli whilst calming the dog and persuading it to remain quiet.
   - *Distraction* as the stimulus happens, through treats, praise, or similar.
   - *Reshaping* via clicker training (a form of operant conditioning) or other means to obtain barking behaviour on command, and then shaping the control to gain command over silence.
3. Seeking professional advice from local organisations, dog trainers, or veterinarians.
4. Use of a mechanical device such as a bark collar. There are several types, all of which use a collar device that produces a response to barking that the dog notices:
   - *Citrus spray ("citronella")* - dogs as a rule do not like citrus. At the least, it is very noticeable and disrupts the pattern through surprise. These collars spray citrus around the dog's muzzle when it barks.
   - *Sonic/ultrasonic (including vibration)* - these collars produce a tone which humans may or may not be able to hear, in response to barking. Over time, the sound becomes annoying or distracting enough to deter barking.
     - o *Electrical* - these collars produce a mild stinging or tingling sensation in response to a bark. It is important that such devices have a failsafe mechanism and shut off after a certain time, to prevent ongoing operation.
     - o *Combination and escalation devices* - many sound and/ or electrical collars have combination or escalation systems. A combination system is one that (for example) uses both sound and spray together. An escalation device is one that uses quiet sounds, or low levels of output, rising gradually until barking ceases. Escalation devices are effective since they "reward" the dog for stopping

sooner by not having "all-or-nothing" action, so the dog can learn to react by stopping before much happens.

Devocalisation

Devocalisation (also known as ventriculocordectomy or vocal cordectomy, and for dogs, debarking or bark softening) is a surgical procedure applied to dogs and cats, where tissue is removed from the animal's vocal cords in order to permanently reduce the volume of their vocalisations. The procedure is outlawed as a form of mutilation in the United Kingdom and all countries that have signed the European Convention for the Protection of Pet Animals and attempts have been made to outlaw the procedure in various parts of the United States.

Procedure and Risks

The procedure may be performed via the animal's mouth, with a portion of the vocal folds removed using a biopsy punch, cautery tool, scissor, or laser. The procedure may also be performed via an incision in the throat and through the larynx, which is a more invasive technique. All devocalisation procedures require general anesthesia. Risks and side effects include negative reaction to anaesthesia, infection, bleeding, and pain. There is also the risk of the removed tissue growing back, or of scar tissue blocking the throat, both requiring further surgeries, though with the incisional technique, the risk of fibrosis is virtually eliminated. Most debarked dogs have a subdued "husky" bark, audible up to 20 metres.

Factors

Chronic, excessive vocalisation may be due to improper socialisation or training, stress, boredom, fear, or frustration. Up to 35% of dog owners report problems with barking, which can cause disputes and legal problems. The practice is more common among some breeds of dog, such as the Shetland Sheepdog (or "Sheltie"), which are known as loud barkers (due to the nature of the environment in which the breed was developed).

National Council on Pet Population and Study Statistics

The National Council on Pet Population Study and Policy (NCPPSP) is a council of ten prominent American animal organisations which studies and addresses statistics on companion animals.

The NCPPSP's *Shelter Statistics Survey* collected data from over 5,000 shelters, The study concluded that neither excessive vocalisation nor general " behaviour problems" were among the top ten reasons companion animals are relinquished at shelters. In a study of 12 shelters reporting behaviours of animals relinquished to shelters as reported by prior caretakers, a majority of relinquished cats and dogs were reported to have "rarely or never" have been too noisy.

Table: *Behaviours of Animals as Reported by Ownersin 12 U.S. Shelters (1995–1996)*

Was too noisy	***Dogs***		***Cats***	
	Number	***Percentage***	***Number***	***Percentage***
Always	99	5.0 %	31	2.4 %
Mostly	179	9.1 %	67	5.2 %
Sometimes	575	29.2 %	242	18.7 %
Rarely/never	1,119	56.7 %	951	73.7 %

Context

Dr. Kathy Gaughan, assistant professor of clinical sciences at Kansas State University's Veterinary Medical Teaching Hospital, points out that "the surgery stops the barking, but it doesn't address why the dog was barking in the first place." Gaughan notes that visitors to her clinic who request debarking are usually looking for a "quick fix". Gaughan states that, commonly, those who seek debarking live in apartments, or have neighbors who complain. Gaughan also counts "breeders with many dogs" among those who most often seek convenience devocalisation. However, Dr. Gaughan does not agree with those who claim the procedure is cruel, stating "Recently, some animal advocates have asserted this surgery is cruel to the animal; some countries have even outlawed the procedure. I do not believe the surgical procedure is cruel; however, failing to address the underlying factors is inappropriate."

Some breeders seek the surgery in order to limit or diminish noise levels for personal reasons ranging from convenience to prevention; some breeders even seek the surgery for puppies prior to going to new homes. However, the fundamental flaw lies within the dog owner's capacity to make appropriate decisions based on their ability to properly provide and care for their breed of choice, which includes taking into consideration living conditions.

Most dogs can be taught to offer more acceptable behaviours in lieu of barking and other noisy behaviours. It is important to remember that dogs may make noise while playing, when excited, stressed, bored, under or over exercised, and for any number of other reasons. Dog owners need to make educated decisions about the breed(s) and breed mixes they choose to bring into their homes by doing research on the history of the breed(s), the breed today, and the necessary requirements for the breed.

Controversy

The subject of convenience devocalisation is controversial.

Opinions of Medical Organisations and Humane Societies

Multiple animal medicine and animal welfare organisations discourage the use of convenience devocalisation, recommending that it only be used as a last resort. However, organisations such as the American Veterinary Medical Association, American Animal Hospital Association and the American Society for the Prevention of Cruelty to Animals, oppose laws that would make devocalisation illegal.

The American Veterinary Medical Association's official position states that "canine devocalisation should only be performed by qualified, licensed veterinarians as a final alternative after behavioural modification efforts to correct excessive vocalisation have failed."

The AVMA's position was later adopted by the American Animal Hospital Association.

The American Society for the Prevention of Cruelty to Animals (ASPCA) recommends that animal caretakers first attempt to address animal behaviour problems with humane behaviour modification techniques and/or with a treatment protocol set up by an animal behaviour specialist. The ASPCA recommends surgery only if behaviour modification techniques have failed, and the animal is at risk of losing its home or its life.

Legal Restriction and Banning

The legality of convenience devocalisation varies by jurisdiction.

United Kingdom

Debarking is specifically prohibited in the UK, along with ear cropping, tail docking, and declawing of cats. By law, convenience devocalisation is considered a form of surgical mutilation.

United States

In the United States, laws vary by state. In the year 2000, anti-debarking legislation was proposed in California, New Jersey, and Ohio. The California and New Jersey bills failed, partially due to opposition from groups who predicted the ban would lead to similar bans on ear cropping and other controversial cosmetic surgical procedures on dogs.

The Ohio bill survived, and was signed into law by Governor Robert Taft in August 2000. However, Ohio Revised Code 955.22 only outlawed debarking of dogs considered "vicious".

In February 2009, 15-year-old Jordan Star of Needham, Massachusetts, filed a bill to outlaw performing convenience devocalisation procedures upon cats and dogs. The bill is co-sponsored by Senator Scott Brown, with the title *Logan's Law*, after a debarked sheepdog. Star said of convenience devocalisation: "To take a voice away from an animal is morally wrong." The bill became state law on April 23, 2010.

Representation

Woof is the conventional representation in the English language of the barking of a dog. As with other examples of onomatopoeia or imitative sounds, other cultures "hear" the dog's barks differently and represent them in their own ways. Some of the equivalents of "woof" in other European and Asian languages are as follows:

- English - woof, woof; ruff, ruff; arf, arf (large dogs and also the sound of sea lions); yap, yap; yip, yip (small dogs), bow wow
- Afrikaans - blaf, blaf; woef, woef; keff, keff (small dogs)
- Albanian - ham, ham
- Arabic - hau, hau; how how
- Armenian -haf, haf
- Basque - au, au; txau, txau (small dogs); zaunk, zaunk (large dogs); jau, jau (old dogs)
- Balinese - kong, kong
- Bengali - gheu, gheu; bhao, bhao
- Bulgarian - bau-bau (áàó-áàó); jaff, jaff (äæàô-äæàô)
- Brazilian Portuguese - au au
- Burmese - woke, woke
- Catalan - bau, bau; bub, bub
- Chinese, Cantonese - wow, wow (jljl)
- Chinese, Mandarin - wang, wang
- Croatian - vau, vau
- Czech - haf, haf; štìk (the bark itself)
- Danish - vov, vuf
- Dutch - waf, waf; woef, woef
- Esperanto - boj, boj

- Estonian - auh, auh
- Finnish - hau, hau; vuh, vuh; rauf, rauf
- French - waouh, waouh; ouahn, ouahn; vaf, vaf; wouf, wouf; wouaf, wouaf; jappe jappe
- German - wuff, wuff; wau, wau; rawrau, rawrau
- Greek - ghav, ghav (ãáâ, ãáâ)
- Hebrew - hav, hav; hau, hau
- Hindi - bow, bow
- Hungarian - vau, vau
- Icelandic - voff, voff
- Indonesian - guk, guk
- Irish - amh, amh
- Italian - bau, bau
- Japanese - wan-wan; kyan-kyan
- Korean - meong, meong (M°M°, pronounced [mŒKmŒK])
- Latvian - vau, vau
- Lithuanian - au, au
- Macedonian - av, av
- Malay - gong, gong ("menggonggong" means barking)
- Marathi - bhu, bhu; bho, bho
- Norwegian - voff, voff or boff
- Persian - vogh, vogh
- Polish - hau, hau
- Portuguese - nasal diphthongtoddler language; cain-cain (whining)
- Romanian - ham, ham; hau, hau
- Russian - gav, gav; tyav, tyav (small dogs)
- Serbian - av, av
- Sinhala - buh, buh
- Slovak - haf, haf; hau, hau
- Slovene - hov, hov
- Spanish - guau-guau; gua, gua; jau, jau
- Swedish - voff, voff; vov, vov; bjäbb, bjäbb
- Tagalog - aw, aw; baw, baw
- Tamil - wal wal

- Thai - pronounced ; pronounced
- Turkish - hev hev; hav, hav
- Ukrainian - (hau, hau); (dzyau, dzyau)
- Urdu - bow bow
- Vietnamese
- Welsh - wff, wff.

Naturally "Barkless" Dog Breeds

Compared to most domestic dogs, the bark of a dingo is short and monosyllabic. During observations, the barking of Australian dingoes was shown to have a relatively small variability; sub-groups of bark types, common among domestic dogs, could not be found. Furthermore, only 5% of the observed vocalisations were made up of barking. Australian dingoes bark only in swooshing noises or in a mixture atonal/ tonal. Also, barking is almost exclusively used for giving warnings. Warn-barking in a homotypical sequence and a kind of "warn-howling" in a heterotypical sequence has also been observed. The bark-howling starts with several barks and than fades into a rising and ebbing howl and is probably, similarly to coughing, used to warn the puppies and members of the pack. Additionally, dingoes emit a sort of "wailing" sound, which they mostly use when approaching a water hole, probably to warn already present dingoes. According to the present state of knowledge, it is not possible to get Australian dingoes to bark more frequently by making them asosciate with other domestic dogs. However Alfred Brehm reported a dingo that completely learned the more "typical" form of barking and knew how to use it, while its brother did not. Whether dingoes bark or bark-howl less frequently in general is not sure.

The now extinct Hare Indian dog of northern Canada was not known to bark in its native homeland, though puppies born in Europe learned how to imitate the barking of other dogs. When hurt or afraid, it howled like a wolf, and when curious, it made a sound described as a growl building up to a howl.

The Basenji of central Africa produces an unusual yodel-like sound, due to its unusually shaped larynx. This trait also gives the Basenji the nickname "Barkless Dog."

Barking in Other Animals

Many animals communicate via various vocalisations. While there is not a precise, consistent and functional acoustic definition for barking, researchers may classify barks according to several criteria. University of Massachusetts Amherst researchers identified volume, pitch, tonality, noise, abrupt onset and pulse duration are amongst the criteria that can be used to define a bark. Besides dogs and wolves, other canines like coyotes and jackals can bark. Their barks are quite similar to those of wolves and dogs. The bark of a dingo is short and monosyllabic.

The warning bark of a fox sounds much like a dog's, but generally the vocalisation of foxes is higher and more drawn out than barks of other canids.

There are also non-canine species with vocalisations that could be described as barking. Because the Muntjac's alarm call resembles a dog's bark, they are sometimes known as Barking Deer. Eared seals are also known to bark. Prairie dogs employ a complex form of communication that involves barks and rhythmic chirps. A wide variety of bird species produce vocalisations that include the canonical features of barking, especially when avoiding predators.

Dog Communication

Dog communication refers to body movements and sounds dogs use to send signals to other dogs, and other animals (usually humans). Dog communication comes in a variety of forms, and is part of the foundation of dog social behaviour. Dogs use certain movements of their bodies and body parts and different vocalisations to express their emotions. There are a number of basic ways a dog can communicate its feelings. These are movements of the ears, eyes, eyebrows, mouth, nose, head, tail, and entire body, as well as barks, growls, whines and whimpers, and howls.

Interpreting Animal Body Language

It is important to note that while many gestures and actions may have common, stereotypical meanings, researchers regularly find that animal communication is often more complex and subtle than previously believed, and that the same gesture may have multiple distinct meanings depending on context and other behaviours. So, generalisations such as "X means Y" are *often*, but not *always* accurate. For example, even a simple tail wag may (depending on context) convey many meanings including:

- Excitement
- Anticipation
- Playfulness
- Contentment/enjoyment
- Happiness, self-confidence.

But also:

- Anxiety
- Uncertainty/apprehension.

There is a simple way to tell the difference though. Just like humans show more happiness on the right side of their face, dogs swing their tail more to the right when they're happy and vice versa. Combined with other body language, in a specific context, many gestures such as yawns and direction of vision all convey the dog's emotions or feeling states. Thus statements that a particular action "means" something, or that the dog is using its body language with the intent to report information to others, should be avoided. According to Daniel C. Dennett of Tufts University in *Consciousness Explained* tells us that there are two basic forms of communication: the unconscious expression of a mental or emotional state, and the intentional act of reporting information. He goes on to say that there are many ways of expressing a mental or emotional state, "but only one way of reporting information, and that is through the use of language, written, spoken, or signed." It's also important to note that most of the body language exhibited by human beings isn't done consciously, with the intent to communicate, so it could be a mistake to believe that dogs intentionally use their body language to report information to others.

Descriptions in this article are therefore best viewed as common generalisations, to which a more experienced observer will be able to add further detail or understanding.

Ethology is a science which studies animal behaviour.

Greeting Ritual

One of the first forms of communication that will be observed is the greeting ritual. When a dog first encounters another dog, a brief assessment of aggression or friendliness is made. If one dog growls or barks, for instance, the encounter will usually end quickly, either by the other dog avoiding the encounter, or by a fight ensuing. If this test is passed, the dogs usually attempt to greet each other. This is done first by sniffing each other's odors. Dogs often sniff each other's anuses simultaneously, and this is the clearest indication of what some in the field believe to be a greeting ritual.

This so-called greeting ritual is said to establish the identities of the dogs by scent, and is the dogs' way of saying 'hello' to each other. Even dogs who know each other very well will sniff one another when they first run into each other on the street. Dogs who live in multiple dog households will also sniff each other from time to time (for instance, if one dog gets up to get a drink of water while another dog is asleep, that dog might go over and sniff his housemate on his way to the water bowl). Dogs who are playing will sometimes get too wound up, stop, shake themselves off, then sniff each other before resuming their play session. If the dogs are satisfied with the encounter (it is not unusual for dogs to take a sudden dislike to each other at this stage), then they may either move on in disinterest, or proceed further in the greeting ritual by showing affection. Affection is shown by some or all of the following: Wagging the tail, licking the face, playful barking, panting, or jumping (including playful jumping on the other dog). Dogs that show affection in this way will usually get along fairly well, and this display can be considered a display of friendship.

Hand-sniff Greeting

Humans can also participate in a greeting ritual with a newly met dog, by bending down in front of (not looming over) or kneeling down to the dog, and slowly but confidently extending the hand to be sniffed in front of and just below the dog's snout. If the dog is timid or has a habit of snapping at strangers, it is best to allow the dog to come sniff your hand, rather than extending it into the dog's space (this can make the dog nervous) while using words of praise in a calm, soothing voice. To limit the chance of getting bitten, keep the hand palm-down with fingers cupped downward or the hand fully closed in a loose fist, making it difficult for the dog to grab hold of a finger in a bite. Be watchful of the dog's demeanor. If you continue praising the dog in a soothing voice in an attempt to calm it after it has just growled or snapped at you, the dog is hearing its bad behaviours get reinforced and is more likely to get frightened and bite, rather than sniff your hand in a friendly manner. Reserve praise and soothing tones for times when the dog is behaving well, and firm, disapproving tones (not yelling) for times when discipline is necessary.

After the dog has completed the hand-sniff, it is possible to proceed to making physical contact by gently petting the dog on its chest or shoulders. Attempting to pet the top of the head can create a nervous response because the movement of the hand toward the head may interrupt the dog's ability to see your eyes, thereby assessing your emotional state. Again, it is possible to get snapped at, so care should be taken not to block the dog's ability to see your eyes. If the dog completes the sniff without snapping or barking, another attempt to pet the dog can be made.

Once the dog allows the affectionate petting, it will more likely only take a quick hand-sniff on the next meeting for the person to attempt petting the dog. Petting can at this time become more playful without risking the dog snapping at the person.

For timid or mildly aggressive dogs, it may not be possible to establish friendship in one greeting ritual. Friendship cannot be forced, and may require repeated attempts over time. Do not look the dog in the eye for this may intimidate them and may snap or growl. Look slightly away, thus seeming less aggressive.

Aggression During the Greeting Ritual

Some breeds of dog have a more suspicious or aggressive temperament by nature and are more difficult or dangerous to approach with the greeting ritual. Dogs that have been physically abused tend to be much more timid and defensive than a well-treated dog, so great care should be taken before trying to perform the greeting ritual with such a dog, as these dogs are more prone to react aggressively. Some dogs are also trained to be aggressive, such as guard dogs.

Dominance and Submission

Dominance and submission are often mistaken to be part of normal social behaviours for dogs. They are not. Wild canines form packs specifically for the purpose of hunting large prey. Evolutionary biologist Raymond Coppinger has noted that wolves that live near garbage dumps, and therefore don't need to hunt large prey, don't form packs. He also states that coyotes, which are more solitary than wolves, sometimes form packs, but only when they need to hunt large prey. In *Dog Language*, biologist Roger Abrantes notes that it's easier for a group of wolves to hunt large prey by working together. So pack formation in canines seems to be a function of prey size more than dominance and submission.

The idea that dogs exhibit dominant and submissive behaviours is based partly on behaviours seen in captive wolves that were culled from various sources, didn't know one another, and weren't able to hunt together. David Mech of the University of Minnesota has been studying wild wolf packs since the 1960s. Mech states that in wild packs "dominance" displays are so rare as to be totally nonexistent. The only time they seem to take place is when a conflict emerges between the pack parents over how to disperse food to the young. The female invariably wins these encounters by acting as "non-threatening" (or submissive) as possible. Rudolph Schenkel was the first biologist to ask the question, if the "submissive" wolf always wins, who's really dominant? Also, since "dominance aggression" in dogs can be treated with anti-anxietal medications, it's more likely that this behaviour is an expression of stress or anxiety, and is not a natural part of the canine social instincts. (Mech (1999) asserted that the significance of dominance relationships within pack society has been overrated, and he argued that wolf packs are best understood as family groups in which a breeding pair "shares leadership in a division of laboursystem in which the breeding female initiates pup care and the breeding male leads in foraging and food provisioning".

Body Movements

Tail: How high or low the tail is held, in relation to how the dog's breed naturally carries its tail, and how it is moved can signify the dog's mood. When the tail is held high, it shows that the dog is alert and aware; the tail between the legs means that the dog is afraid or frightened. If the fur on the tail is also bristled[pokey feeling], the dog is saying it is willing to defend itself or pups.

Small, slow wags of the tail say the dog is questioning things around the environment it is in. Either it is not sure whether it is the target dog or the person around them is friendly, or it is not sure what is going on or what is expected of it to do. Large, fast wags of the tail may be a sign of a happy, excited, or an energetic dog, but can also signal aggression. A large percentage of the victims of dog bites are bitten while the dog is wagging its tail.

Dogs are said to exhibit a left-right asymmetry of the tail when interacting with strangers, and will show the opposite, right-left motion with people and dogs they know.

Aggressive/ Violent

These dogs are showing the major signs of aggression, but they are not fighting, so body language should not be inferred on isolated signs. When a dog's lips curl back this shows that the dog has a strong urge to bite. This is an unconscious reflex, designed to get the soft flesh of the lips away from the teeth before the dog bites, and is often misinterpreted as a way of communicating aggressive intent. For example, many dogs will curl their lips back into a "snarl" when they take a cookie or bone.

Ear position relates the dog's level of attention, and reaction, to a situation or animal. Erect ears facing forward means the dog is very attentive. They lay their ears back for the sounds surrounding them and also when in a submissive state. Dogs with drop ears, like Beagles, can't use these signals very well, as the signals first developed in wolves, whose ears are pricked. Wolf-like dogs (such as the Samoyed or Husky) will, when content and happy, often hold their ears in a horizontal position but still forward. This has been referred to as the "wolf smile". Mouth expressions can provide information about the dog's mood. When a dog wants to be left alone, it might yawn (although yawning also might indicate sleepiness, confusion, or stress) or start licking its mouth without the presence of any food.

When a dog is happy or wants to play, it might pant with lips relaxed, covering the teeth and with what sometimes appears to be a happy expression (it might appear as a smile to some observers) or with the mouth open. Mouth expressions that indicate aggression include the snarl, with lips retracting to expose the teeth, although some dogs also use this during play. However, some dogs will pull back their "top lips" in what looks like an aggressive way, when they are excited or happy. For example a dog prone to "smiling" may do so in greeting to a much loved owner and this should not be punished lest the dog become less affectionate and more withdrawn.

It's important to look at the dog's whole body and not just the mouth or tail before deciding what the dog is feeling. What appears initially as aggression might be an invitation to play, or vice-versa.

Tongue (Licking)

A very common form of communication is for a dog to lick another dog, or a person. Dogs lick other dogs' faces and mouths when they greet each other to indicate friendliness. Dogs like to lick human skin not only for the salt from the sweat, but also as a form of greeting, such as by briefly licking a person's hand after sniffing it. Licking is also used as a social bonding analogous to primate social grooming and stroking. This can indicate intimacy. Such licking is longer and slower, as compared to the brief licking of faces during a greeting.

Eyes and Eyebrows

While dogs don't have actual eyebrows, they do have a distinctive ridge above their eyes, and some breeds, like the Labrador Retriever, Rottweiler, Bernese Mountain Dog, German Shepherd, and Doberman have markings there. A dog's eyebrow movements usually express a similar emotion to that of a human's eyebrow movements. Raised eyebrows suggest interest, lowered brows suggest uncertainty or mild anger, and one eyebrow up suggests bewilderment. Eyes narrowed to slits indicate affection for the person or animal the dog is looking at.

Feet and Legs

Although a dog's feet lack the dexterity of human hands, a dog can use them as an avenue of communication. A dog might stamp its feet, alternating its left and right front legs, while its back legs are still. This occurs when the dog is excited, wants something, or wants its owner's attention. Pointers tend to tuck one front leg up when they sense game nearby. This behaviour is not communicative so much as the dog exhibiting a fixed-action pattern called "the eye stalk." It is also common for dogs to paw or scratch for objects they desire. Many dogs are trained to mimic a human handshake, offering a paw to a human stooping down and offering their own hand in exchange.

Head

The leaning of a dog's head to the right or to the left often indicates curiosity and/or a sound it has not heard before. This, however, may also be a sign of recognition to a familiar word.

If the dog's head is held high with its neck craning forward, it is showing interest, although, it could also mean an aggressive mood if other body language is present.

Some adult dogs that were not properly raised have been known to challenge their owners for alpha position. One of the signs, though this is rarely seen in dogs, involves the dog slightly lowering its head while standing tall with its eyes fixed upward at the owner or any human beings they are about to challenge (start a fight with). This behaviour is extremely rare and usually occurs with dogs that have been severely neglected or in some cases, abused. This can also be dangerous and sometimes fatal if no action is taken immediately. However, this behaviour is preventable if owners avoid being neglectful or abusive to their dogs.

Canine Reproduction

Canine reproduction is a social behaviour of the domestic dog.

Canine Sexual Anatomy and Development

General: In breeding, the female sex of the domestic canine is often referred to as a bitch and the male of the same species is referred to as a stud. The dog chosen for breeding is the stud. In domestic canines, sexual maturity (puberty) occurs between the ages of 6 to 12 months for both males and females, although this can be delayed until up to two years of age for some large breeds.

Pregnancy is possible as soon as the first estrus cycle, but breeding is not recommended prior to the second cycle. As with other domesticated species, domestication has selectively bred for higher libido, and earlier and more frequent breeding cycles in dogs than in their wild ancestors.

Male

Like most males of any species, the development of secondary sexual characteristics is dependent on the production of testosterone by the testes. Secondary sex characteristics include increased muscle mass, penile enlargement, anal hypertrophy, and development of a gland at the base of the penis, called the bulbus glandis. Testosterone levels are also responsible for the sex drive, which is not cyclical. Males are receptive to mating at any time, even if the female is not in estrus.

Female

The anatomy of female dogs is similar to that of many other mammals. They have two ovaries located caudal to the kidneys in the abdominal cavity. They have a bicornuate (or two-horned) uterus which ends as the cervix, a short canal which connects to the vagina. The cervix is muscular with fibrous tissue support, and its opening closes in a stellate pattern. The vagina opens into an enlarged space called the vestibule which also receives the opening of the urethra. The vestibule makes a steep downward turn exits within the vulva. Females have a clitoris just inside the vulva. Bitches often have a vaginal stricture, which is a remnant of where the vagina and vestibule fused together during embryonic development. This structure is often asymptomatic and is broken down during mating.

As with most domesticated species, one of the first and strongest effects seen from selective breeding is selection for cooperation with the breeding process as directed by humans. In domestic dogs, one of the behaviours that is noted is the abolition of the pair bond seen in wild canines. The ability of the female domestic dog to come into estrus at any time of the year and usually twice a year is also valued. The amount of time between cycles varies greatly among individual dogs, but a particular dog's cycle tends to be consistent through her life. Conversely, undomesticated canine species experience estrus once a year, typically in late winter.

Most female dogs come into season for the first time between 6 and 12 months, although some larger breeds delay until as late as 2 years. Like most mammals, the age that a female first comes into season is mostly a function of her current body weight as a proportion of her body weight when fully mature. They then experience estrus about every seven months until old age. Female dogs do not experience menopause, although their cycles will become irregular and fertility becomes unpredictable as they become older. Dogs over around 7 or 8 years are not usually considered appropriate for breeding, but can still remain fertile.

The Reproductive Cycle

Female Cycle: The average length of the reproductive cycle for females is 6 months. Females reach sexual maturity (puberty) between 8 and 18 months of age. There is a tremendous variability in the maturation age between breeds, and even within a breed of dog.

The first stage of the reproductive cycle is proestrus, in which eggs in the ovaries begin to mature and estrogen levels begin to rise. During this stage males are attracted to non-receptive females. Initially, the vulva lips will swell up and become pliable, there are small amounts of bloody vaginal discharge, and signs of frequent urination and restlessness. Proestrus generally lasts 9 days.

Estrus is the next stage, in which estrogen levels are high, mature eggs are released from ovaries, and the females mentally and physically become receptive to copulation. It is only during estrus that copulation will result in pregnancy. During proestrus and estrus, females may have a clear to bloody discharge. This stage is also known as "heat." The length of these cycles varies greatly between individuals. Proestrus and estrus can last anywhere from 5 days to 21 days.

Diestrus is the period following mating. Diestrus lasts approximately 56 to 58 days in a pregnant female, and 60 to 100 days in a non-pregnant female. During both of these periods, progesterone levels are high. Because the hormonal profile of a pregnant female and a female in diestrus are the same, sometimes a non-pregnant female will go through a period of pseudo-pregnancy. At that time she may gain weight, have mammary gland development, produce milk, and exhibit nesting behaviours.

Anestrus is the period of reproductive quiescence. The female has no attraction to or from the male. Anestrus generally lasts four to five months.

Copulation

Copulation between canines involves a male canine mounting the female from behind, as with most tetrapods, a position known informally as doggy style. When a male canine is interested in mounting a female, he will sniff the female's vulva. If the female is unreceptive, she may sit or lie down, snap, move away, or otherwise be uncooperative or not allow the mating. If the female is receptive to mating, she will stand still and hold her tail to the side (called "flagging"). The male will often continue examining the female's rear, before mounting her from behind whilst attempting penetration with his penis.

At the time of penetration, the penis is not erect; it is slender and is only able to penetrate the female because of the presence of the small bone inside it, known as the baculum. When the male achieves penetration, he will often hold the female tighter and thrust faster, and it is at this point that the male's penis expands. Unlike human sexual intercourse, where human males become erect first and then enter the female, canine copulation involves the canine male first entering the female, and then swelling and becoming erect.

Male canids are the only animals that have a bulb or locking bulbus glandis, a spherical area of erectile tissue at the base of the penis. During copulation, after the male's penis is fully inside the female's vagina, the bulbus glandis becomes engorged with blood. Then, when the female's vagina contracts, the penis is locked inside the female. This is known as *tying* or alternatively *knotting*. While characteristic of mating in most canids, the copulatory tie has been reported to be absent or very brief (less than one minute) in the African Wild Dog, possibly due to the abundance of large predators in its environment.

When the penis is locked into the vagina by the bulbus glandis (when the stud is "tied"), the male will usually lift a leg and swing it over the female's back while turning around. The two stand with their hind ends touching and the penis locked inside the vagina while ejaculation occurs, decreasing leakage of semen from the vagina. After some time, typically between 5 and 20 minutes (but sometimes longer), the bulbus glandis disengorges, allowing the mates to separate. Virgin dogs can become quite distressed at finding themselves unable to separate during their first copulation, and may try to pull away or run. Dog breeders often suggest it is appropriate for those involved to help calm and settle the mating dogs once this stage is reached, if they show anxiety, through until eventual separation. Similar canine mounting behaviour (sometimes with pelvic thrusting) is also used by dominant canines of both sexes. Dominance mounting, with or without thrusting, should not be confused with copulatory mounting, in which the thrusting is short term until a "tie" is achieved. In dominance mounting, the dominant dog is merely trying to get its head above the subordinate dog. If a male dog does this, one may notice that the genitals are aimed *over* the subordinates tail rather than *under* it. Unsheathing of the male genitals is a sign that the dominant dog has "won the fight" of dominance.

Gestation and Litters

Dogs bear their litters roughly 9 weeks after fertilization, although the length of gestation can vary from 59 to 65 days, with 63 days being the average. A general rule of thumb is that a mammal will produce half as many offspring as the number of teats on the mother. This rule is altered in domesticated animals since larger litters are often favoured for economic reasons and in dogs, particularly, the great range of sizes and shapes plays a role in how many healthy puppies a female can carry. An average litter consists of about six puppies, though this number may vary widely based on the breed of dog. Toy dogs generally produce from one to four puppies in each litter, while much larger breeds may average as many as 14 pups in each litter. The number of puppies also varies with the mother's age and health, the father's sperm count, the timing of the breeding, and many other factors.

Some veterinarians say that mating a stud and bitch every other day is the way to a full litter. Others will say that mating every day is much better. This reduces the chance of having one puppy being born prematurely or much later than the other pups of the litter.

Some breeds have been developed to emphasize certain physical traits beyond the point at which they can safely bear litters on their own. For example, an English Bulldog has wide shoulders and narrow hips, thus the breed requires artificial insemination and, later, a cesarean section for giving birth. Since a mother can provide nutrients and care to only a limited number of offspring, humans must assist in the care and feeding when the litter exceeds approximately eight puppies.

Clinical Issues

Owners of intact bitches should be aware of pyometra, an acute infection similar in effect to appendicitis which can affect any intact bitch, and which is always very serious and often fatal. The period after heat, when the cervix is closing, is a high risk period for this disease and close attention should be paid for any signs of fever, lethargy, loss of appetite, excessive thirst, or any other indicators of sickness. Pyometra should be treated as a medical emergency if suspected. Contrary to common belief, pyometra can strike any intact bitch, whether she has been bred or not, and whether it is her first or tenth season, although it is more common as dogs become older.

Dog Breeding

Timing: Breeders and veterinarians have various methods for determining the best time to breed a dog. Breeders rely on both the visible signs of estrus and modern testing methods when they are trying to decide the best time to breed. Visible signs of estrus (or heat) include vaginal bleeding, vulvar swelling and the female dog responding to a male dog by standing and "freezing" to allow the male dog to mount her. The three most commonly used medical methods for determining estrus are: blood progesterone, blood luetinizing hormone(LH), and a vaginal smear. The first two test for the level of those hormones in the blood, and a "smear" checks the amount of cornified epithelial cells present in the vaginal secretions.

Many breeders prefer the LH and progesterone tests because they are highly accurate while the smear method can be unreliable (especially when done by a vet who has limited experience performing this test). However the smear method is still widely used because it is much cheaper, and it's less stressful for the dog because no blood has to be drawn. Before the advent of modern testing a breeder might mate the two dogs every two to three days during the female's estrus to ensure that fertilization had occurred. Today, most breeders will track a female's cycle (usually by progesterone blood tests) and breed her only during a 3-day window when she is most fertile and most receptive to the male. A female dog allowed to roam freely during estrus can end up producing puppies from multiple fathers. Breeders might occasionally breed a female to more than one desirable male for various reasons, in which case the only reliable way to determine parentage of the puppies is through DNA testing of the pups and mother.

Procedure

Natural breeding can be easily accomplished between most dogs as long as the bitch is in estrus during the attempt. Simply placing both dogs in the same environment for a few days will usually result in a pregnancy. If, due to breed characteristics or the stud only being available for a short time, natural breeding is not desirable, artificial insemination can be used. This is often done at a veterinarian's office. An artificial vagina is prepared, which is a conical thin latex sleeve ending in a sterile collection tube. The inside of the latex sleeve is lightly lubricated. The male is allowed to sniff a female in estrus. Experienced studs cooperate readily in the process. New studs often require encouragement in the form of manual stimulation. Generally the male will mount the female, and the collector quickly directs his penis into the latex sleeve.

The male ejaculates and the semen is collected in the tube. The semen is then drawn up into a long thin pipette. The pipette is threaded through the bitch's cervix and the semen deposited in her uterus. There is ongoing research into techniques for chilling or freezing canine semen. Currently, for high quality breedings, one of the mates is often flown to the location of the other for this procedure.

Amateur Breeding

Amateur breeding is also referred to by some as "backyard breeding" and refer to those specifically who plan on selling the offspring on a regular basis for income. Natural breeding can also take place in private homes. Purebred dog breeding arose in response to jobs for which a specific type of dog was needed, whether hunting, guarding, herding, or other work and were also bred simply for their physical attractiveness and appearance as well. Most of the work qualities no longer serve any useful purpose in current American society. Different qualities were needed for these various jobs, such as a soft mouth for retrievers or a herding instinct for sheep dogs. In order to fix the genetic breed features so that two dogs of the same breed could reproduce themselves, all breeds have historically started with inbreeding. Those fixed qualities are today referred to as the breed standard. Standards are set and approved by each national breed club. Many standards are quite specific and not open to interpretation, such as height, ear type (prick versus drop), tail, coat, and other conformation features. Other aspects of each breed may be open to interpretation. Breeders may focus on conformation, temperament, or working qualities, such as hunting, pointing, herding, or rescue. Currently, the main goals for most breeders are to improve their chosen breed for show as well as producing puppies for sale as pets. Many commonly advertise their dogs on the internet,selling and shipping, often by plane, for pet and show purposes.

Professional breeders are not only aware of their breed standards, but some attempt to improve their own line by either linebreeding or by introducing new lines. Linebreeding or inbreeding can cause two main problems: the doubling up of disease-causing recessive alleles, causing the offspring to have health problems, and the loss of vigour. Dogs whose inbreeding coefficient is higher than 6.25% have shorter lifespans and are more prone to disease. Professional breeders may test for genetic defects before breeding for such things as hip dysplasia, various eye conditions, thyroid problems, and epilepsy.

Cross Breeding

Designer breed dogs are a Mixed-breed dog that are intentionally bred from parents of two established breeds. Studies have shown that cross-bred dogs have a number of desirable reproductive traits. Scott and Fuller found that cross-bred dogs were superior mothers compared to purebred mothers, producing more milk and giving better care. These advantages led to a decreased mortality in the offspring of cross-bred dogs.It takes 2 months for the puppy to be born.

Chapter 3

Elephant Behaviour

Elephants are large land mammals in two extant genera of the family Elephantidae: *Elephas* and *Loxodonta*, with the third genus *Mammuthus* extinct. Three species of elephant are recognised: the African bush elephant, the African forest elephant and the Indian or Asian elephant; although some group the two African species into one and some researchers also postulate the existence of a fourth species in West Africa. All other species and genera of Elephantidae are extinct. Most have been extinct since the last ice age, although dwarf forms of mammoths might have survived as late as 2,000 BCE. Elephants and other Elephantidae were once classified with other thick-skinned animals in a now invalid order, Pachydermata.

Figure: *African elephant* *Asian elephant*

Elephants are the largest living land animals on Earth today. The elephant's gestation period is 22 months, the longest of any land animal. At birth, an elephant calf typically weighs 105 kilograms (230 lb). They typically live for 50 to 70 years, but the oldest recorded elephant lived for 82 years.

The largest elephant ever recorded was shot in Angola in 1955. This male weighed about 24,000 lb (11,000 kg), with a shoulder height of 3.96 metres (13.0 ft), a metre (yard) taller than the average male African elephant. The smallest elephants, about the size of a calf or a large pig, were a prehistoric species that lived on the island of Crete during the Pleistocene epoch.

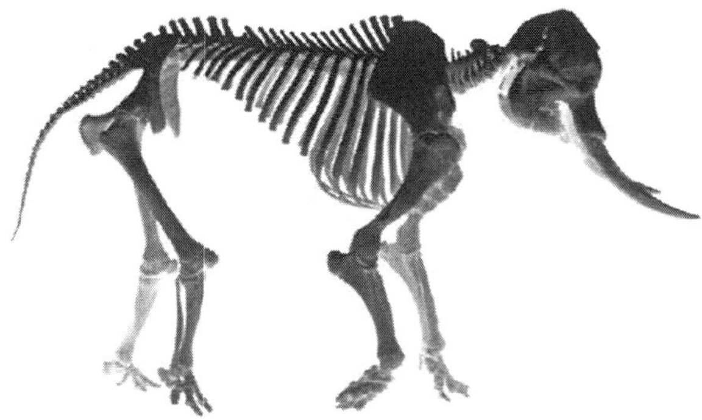

Figure: *African elephant skeleton*

Elephants are a symbol of wisdom in Asian cultures and are famed for their memory and intelligence, where their intelligence level is thought to be equal to that of dolphins and primates. Aristotle once said the elephant was "the beast which passeth all others in wit and mind." The word "elephant" has its origins in the Greek ëÝöáò, meaning "ivory" or "elephant". According to observations, healthy adult elephants have no natural predators, although lions may take calves or weak individuals. They are, however, threatened by human intrusion and poaching.

Etymology

Olifant and its variations (ex. *oliphant, olyphant*) are archaic spellings of *elephant*. Aside from elephants, the word has been used to refer to ivory, elephant tusks, musical horns made of elephant tusks, or a musical instrument resembling such horns. It appears in Middle English as *olifant* or *olifaunt*, and was borrowed from Medieval French *olifanz*. The French word owes something to both Old High German *olbenta* "camel", and to Latin *elephantus* "elephant", a word of Greek origin. OHG *olbenta* is a word of old Germanic origin; cf. Gothic *ulbandus* also meaning "camel". But the form of the OHG and Gothic words suggests it is also a borrowing, perhaps indeed directly or indirectly from Greek "ëÝöáò" (*elephas*), which in Homer only meant "ivory", but from Herodotus on the word also referred to the animal. The earliest attested form of the word is the Mycenaean Greek *e-re-pa-to*, written in Linear B syllabic script.

Taxonomy and Evolution

The African elephant genus contains two or, arguably, three living species; whereas the Asian elephant species is the only surviving member of the Asian elephant genus, but can be divided into four subspecies. The African and the Asian elephants diverged from a common ancestor some 7.6 million years ago.

African Elephant

The Elephants of the genus Loxodonta, known collectively as African elephants, are currently found in 37 countries in Africa.

African elephants are distinguished from Asian elephants in several ways, the most noticeable being their much larger ears. Also, the African elephant is typically larger than the Asian elephant and has a concave back. In Asian elephants, only males have tusks, but both males and females of African elephants have tusks and are usually less hairy than their Asian cousins.

African elephants have traditionally been classified as a single species comprising two distinct subspecies, namely the savanna elephant (*Loxodonta africana africana*) and the forest elephant (*Loxodonta africana cyclotis*), but recent DNA analysis suggests that these may actually constitute distinct species. This split is not universally accepted by experts. A third species of African elephant has also been proposed.

The authors of an analysis of nuclear DNA extracted from "African savanna elephant, African forest elephant, Asian elephant, the extinct American mastodon, and the woolly mammoth" concluded in 2010 that African savanna and forest elephants are indeed separate species:

We unequivocally establish that the Asian elephant is the sister species to the woolly mammoth. A surprising finding from our study is that the divergence of African savanna and forest elephants—which some have argued to be two populations of the same species—is about as ancient as the divergence of Asian elephants and mammoths. Given their ancient divergence, we conclude that African savanna and forest elephants should be classified as two distinct species.This reclassification has implications for conservation. If there are two separate species, each will be less abundant (particularly the rarer) and could be more endangered than a more numerous and wide-ranging single species. There is also a potential danger that if the forest elephant is not explicitly listed as an endangered species, poachers and smugglers might be able to evade the law forbidding trade in endangered animals and their products.

The forest elephant and the savanna elephant can hybridise (interbreed), though their preferences for different terrains reduce such opportunities. As the African elephant has only recently been recognised to comprise two separate species, groups of captive elephants have not been comprehensively classified and some could well be hybrids.

Under the new two species classification, *Loxodonta africana* refers specifically to the savanna elephant, the largest of all elephants. It is the largest land animal, with males standing 3.2 metres (10 ft) to 4 metres (13 ft) at the shoulder and weighing 3,500 kilograms (7,700 lb) up to a reported 12,000 kilograms (26,000 lb). The female is smaller, standing about 3 metres (9.8 ft) at the shoulder. Most often, savanna elephants are found in open grasslands, marshes, and lakeshores. They range over much of the savanna zone south of the Sahara.

The other putative species, the forest elephant (*Loxodonta cyclotis*), is usually smaller and rounder, and its tusks thinner and straighter compared with the savanna elephant. The forest elephant can weigh up to 4,500 kilograms (9,900 lb) and stand about 3 metres (10 ft) tall. Much less is known about these animals than their savanna cousins, because environmental and political obstacles make them difficult to study. Normally, they inhabit the dense African rain forests of central and western Africa, although occasionally they roam the edges of forests, thus overlapping the savanna elephant home ranges and hybridising. In 1979, Iain Douglas-Hamilton estimated the continental population of African elephants at around 1.3 million animals. This estimate is controversial and is believed to be a gross overestimate, but it is very widely cited and has become a *de facto* baseline that continues to be incorrectly used to quantify downward population trends in the species. Through the 1980s, *Loxodonta* received worldwide attention due to the dwindling numbers of major populations in East Africa, largely as a result of poaching. According to IUCN's African Elephant Status Report 2007, there are between 470,000 and 690,000 African elephants in the wild. Although this estimate only covers about half of the total elephant range, experts do not believe the true figure to be much higher, as it is unlikely that large populations remain to be discovered. By far, the largest populations are now found in southern and eastern Africa, which together account for the majority of the continental population. According to a recent analysis by IUCN experts, most major populations in eastern and southern Africa are stable or have been steadily increasing since the mid-1990s, at an average rate of 4.5% per year.

Elephant populations in West Africa, on the other hand, are generally small and fragmented, and only account for a small proportion of the continental total. Much uncertainty remains as to the size of the elephant population in central Africa, where the prevalence of forest makes population surveys difficult, but poaching for ivory and bushmeat is believed to be intense through much of the region. South African elephant population more than doubled, rising from 8,000 to over 20,000, in the thirteen years after a 1995 ban on the trade in elephant ivory. The ban on the ivory trade in southern Africa (but not elsewhere) was lifted in February 2008, sparking controversy among environmental groups.

Asian Elephant

The Asian elephant, *Elephas maximus*, is smaller than the African. It has smaller ears, and typically, only the males have large external tusks.

Figure: *A decorated Indian elephant in Jaipur, India.*

The world population of Asian elephants—also called Indian elephants—is estimated to be around 60,000, about a tenth of the number of African elephants. More precisely, it is estimated that there are between 38,000 and 53,000 wild elephants and between 14,500 and 15,300 domesticated elephants in Asia, with perhaps another 1,000 scattered around zoos in the rest of the world. The Asian elephants' decline has possibly been more gradual than the African and caused primarily by poaching and habitat destruction by human encroachment.

Several subspecies of *Elephas maximus* have been identified, using morphometric data and molecular markers. *Elephas maximus maximus* (Sri Lankan elephant) is found only on the island of Sri Lanka. It is the largest of the Asians. There are an estimated 3,000–4,500 members of this subspecies left today in the wild, although no accurate census has been carried out recently. Large males can weigh upward to 5,400 kg (12,000 lb) and stand over 3.4 m (11 ft) tall. Sri Lankan males have very large cranial bulges, and both sexes have more areas of depigmentation than other Asians. Typically, their ears, face, trunk, and belly have large concentrations of pink-speckled skin. There is an orphanage for elephants in Pinnawala, Sri Lanka, which plays a large role in protecting the Sri Lankan elephant from extinction.

Elephas maximus indicus (Indian elephant) makes up the bulk of the Asian elephant population. Numbering approximately 36,000, these elephants are lighter grey in colour, with depigmentation only on the ears and trunk. Large males will ordinarily weigh only about 5,000 kg (11,000 lb), but are as tall as the Sri Lankan. The mainland Asian can be found in 11 Asian countries, from India to Indonesia. They prefer forested areas and transitional zones, between forests and grasslands, where greater food variety is available.

The Sumatran elephant, *Elephas maximus sumatranus*, found only on Sumatra, is smaller than the Indian elephant. Population estimates for this group range from 2,100 to 3,000 individuals. It is very light grey in colour and has less depigmentation than the other Asians, with pink spots only on the ears. Mature Sumatrans will usually only measure 1.7–2.6 m (5.6–8.5 ft) at the shoulder and weigh less than 3,000 kg (6,600 lb). It is considerably smaller than its other Asian (and African) cousins and exists only on the island of Sumatra, usually in forested regions and partially wooded habitats.

In 2003, a further subspecies was identified on Borneo. Named the Borneo pygmy elephant, it is smaller and tamer than any other Asian elephants. It also has relatively larger ears, longer tail and straighter tusks.

Physical Characteristics

Trunk

The proboscis, or trunk, is a fusion of the nose and upper lip, elongated and specialised to become the elephant's most important and versatile appendage. African elephants are equipped with two fingerlike projections at the tip of their trunk, while Asians have only one.

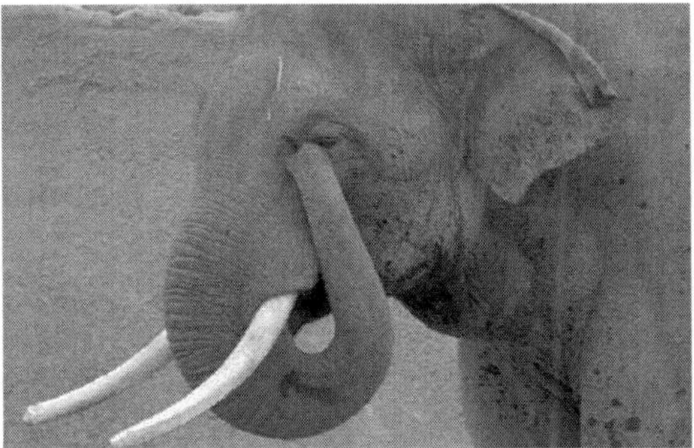

Figure: *An elephant can use its trunk for a variety of purposes. This one is wiping its eye.*

The elephant's trunk is sensitive enough to pick up a single blade of grass, yet strong enough to rip the branches off a tree.

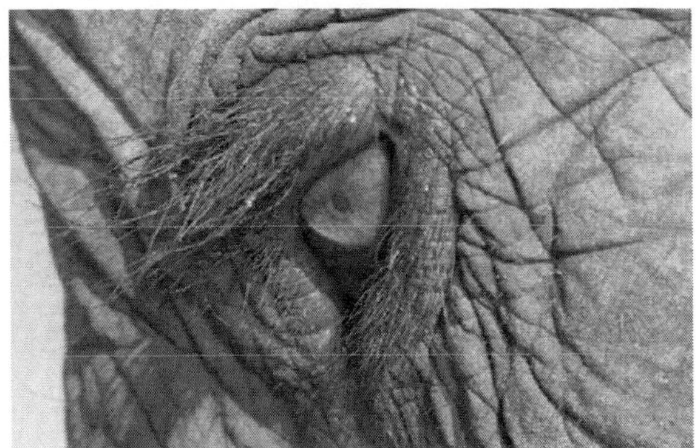

Figure: *Eye of an Asian elephant.*

Most herbivores (plant eaters, like the elephant) possess teeth adapted for cutting and tearing off plant materials. However, except for the very young or infirm, elephants always use their trunks to tear up their food and then place it in their mouths. They will graze on grass or reach up into trees to grasp leaves, fruit, or entire branches. If the desired food item is too high up, the elephant will wrap its trunk around the tree or branch and shake its food loose or sometimes simply knock the tree down altogether.

The trunk is also used for drinking. Elephants suck water up into the trunk—up to 14 litres (15 quarts) at a time—and then blow it into their mouths. Elephants also suck up water to spray on their bodies during bathing. On top of this watery coating, the animals will then spray dirt and mud, which dries and acts as a protective sunscreen.

When swimming, the trunk makes an excellent snorkel. This appendage also plays a key role in many social interactions.

Familiar elephants will greet each other by entwining their trunks, much like a handshake. They also use them while play-wrestling, caressing during courtship and mother-child interactions, and for dominance displays; a raised trunk can be a warning or threat, while a lowered trunk can be a sign of submission. Elephants can defend themselves very well by flailing their trunks at unwanted intruders or by grasping and flinging them.

An elephant also relies on its trunk for its highly developed sense of smell. By raising the trunk up in the air and swiveling it from side to side, like a periscope, it can determine the location of friends, enemies, and food sources.

Some elephants have been afflicted by floppy trunk syndrome.

Tusks

The tusks of an elephant are its second upper incisors. Tusks grow continuously; an adult male's tusks grow about 18 cm (7 in) a year. Tusks are used to dig for water, salt, and roots; to debark trees to eat the bark; to dig into baobab trees to get at the pulp inside; and to move trees and branches when clearing a path. In addition, they are used for marking trees to establish territory, and occasionally as weapons.

Like humans who are typically right- or left-handed, elephants are usually right- or left-tusked. The dominant tusk, called the master tusk, is generally shorter and more rounded at the tip from wear. Both male and female African elephants have large tusks that can reach over 3 m (10 ft) in length and weigh over 90 kg (200 lb). In the Asian species, only the males have large tusks. Female Asians have tusks which are very small or absent altogether. Asian males can have tusks as long as the much larger Africans, but they are usually much slimmer and lighter; the heaviest recorded is 39 kg (86 lb).

The tusk of both species is mostly made of calcium phosphate in the form of apatite. As a piece of living tissue, it is relatively soft (compared with other minerals such as rock), and the tusk, also known as ivory, is strongly favoured by artists for its carvability. The desire for elephant ivory has been one of the major factors in the reduction of the world's elephant population.

Some extinct relatives of elephants had tusks in their lower jaws in addition to their upper jaws, such as *Gomphotherium*, or only in their lower jaws, such as *Deinotherium*.

Teeth

Elephants' teeth are very different from those of most other mammals. Over their lives they usually have 28 teeth. These are:

- The two upper second incisors: these are the tusks.
- The milk precursors of the tusks.
- 12 premolars, 3 in each side of each jaw.
- 12 molars, 3 in each side of each jaw.

Unlike most mammals, which grow baby teeth and then replace them with a permanent set of adult teeth, elephants have cycles of tooth rotation throughout their entire lives. The tusks have milk precursors, which fall out quickly and the adult tusks are in place by one year of age, but the chewing teeth are replaced five or, very rarely, six times in an elephant's lifetime.

Only four chewing teeth (premolars and/or molars), one on each side of each jaw, are in primary use at any given time (or two, as one replaces the other at each location). Adult teeth do not replace milk teeth by emerging from the jaws vertically as human teeth do. Instead, new teeth grow in at the back of the mouth, pushing older teeth toward the front, where the latter break off in pieces until they are gone. In African elephants, the first two sets of chewing teeth (premolars) are in place when an elephant is born. The first chewing tooth on each side in each jaw falls out when the elephant is about two years old. The second set of chewing teeth falls out when the elephant is about six years old. The third set is lost at 13 to 15 years of age, and set four lasts to approximately 28 years of age. The fifth set of chewing teeth (molars) lasts until the elephant is in its early 40s. The sixth (and usually final) set must last the elephant the rest of its life. If an elephant lives to more than 60 years of age, the last set of molars is worn to stumps, and it can no longer feed properly. Moss reports a female elephant in its sixties whose final set of molars were worn smooth and about one-quarter of their original size and who survived "with extra chewing and longer feeding bouts." Abscesses of chewing teeth, as well as of tusks and jaws, are common in elephants, and may lead to premature death.

Tusks in the lower jaw are also second incisors. These grew out large in *Deinotherium* and some mastodons, but in modern elephants they disappear early without erupting.

Skin

Elephants are colloquially called *pachyderms* (from their original scientific classification), which means thick-skinned animals. An elephant's skin is extremely tough around most parts of its body and measures about 2.5 centimetres (1.0 in) thick. However, the skin around the mouth and inside of the ear is considerably thinner.

Normally, the skin of an Asian is covered with more hair than its African counterpart. This is most noticeable in the young. Asian calves are usually covered with a thick coat of brownish red fuzz. As they get older, this hair darkens and becomes more sparse, but it will always remain on their heads and tails.

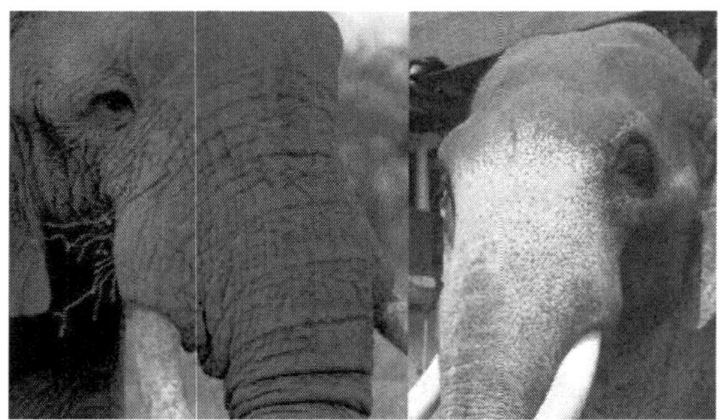

Figure: *Skin of an African (left) and Asian (right) elephant.*

The species of elephants are typically greyish in colour, but the Africans very often appear brown or reddish from wallowing in mud holes of coloured soil. Wallowing is an important behaviour in elephant society. Not only is it important for socialisation, but the mud acts as a sunscreen, protecting its skin from harsh ultraviolet radiation. Although tough, an elephant's skin is very sensitive.

Without regular mud baths to protect it from burning, as well as from insect bites and moisture loss, an elephant's skin would suffer serious damage. After bathing, the elephant will usually use its trunk to blow soil on its body to help dry and bake on its new protective coat. As elephants are limited to smaller and smaller areas, there is less water available, and local herds will often come too close over the right to use these limited resources.

Wallowing also aids the skin in regulating body temperatures. Elephants have difficulty in releasing heat through the skin because, in proportion to their body size, they have very little of it. The ratio of an elephant's mass to the surface area of its skin is many times that of a human. Elephants have even been observed lifting up their legs to expose the soles of their feet, presumably in an effort to expose more skin to the air. Since wild elephants live in very hot climates, they must have other means of getting rid of excess heat.

Legs and feet

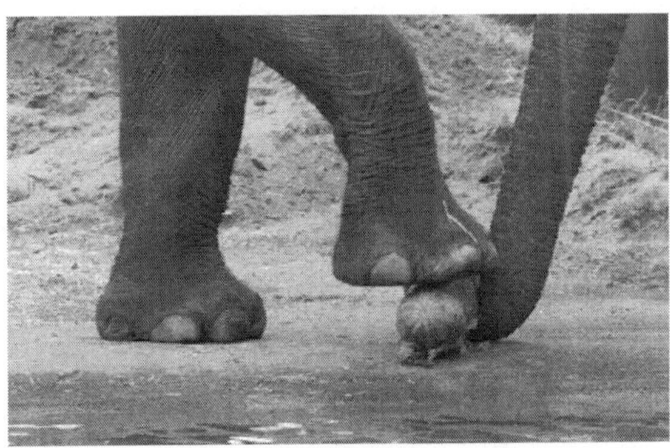

Figure: *Elephant using its feet to crush a watermelon before eating it.*

An elephant's legs are roughly shaped like columns or pillars, as they must be to support its bulk. The elephant needs less muscular power to stand because of its straight legs and large padded feet. For this reason, an elephant can stand for very long periods of time without tiring. In fact, African elephants rarely lie down unless they are sick or wounded. Indian elephants, in contrast, lie down frequently.

The feet of an elephant are nearly round. African elephants have three nails on each hind foot, and four on each front foot. Indian elephants have four nails on each hind foot and five on each front foot. Beneath the bones of the foot is a tough, gelatinous material that acts as a cushion or shock absorber. Under the elephant's weight, the foot swells, but it gets smaller when the weight is removed. An elephant can sink deep into mud, but can pull its legs out readily because its feet become smaller when they are lifted.

Elephants swim well, but cannot trot, jump, or gallop. They do have two gaits: a walk and a faster gait that is similar to running.

In walking, the legs act as pendulums, with the hips and shoulders rising and falling while the foot is planted on the ground. With no "aerial phase", the faster gait does not meet all the criteria of running, as elephants always have at least one foot on the ground. However, an elephant moving fast uses its legs much like other running animals, with the hips and shoulders falling and then rising while the feet are on the ground. In this gait, an elephant will have three feet off the ground at one time. As both of the hind feet and both of the front feet are off the ground at the same time, this gait has been likened to the hind legs and the front legs taking turns running. Tests at the Thai Elephant Conservation Centre are reported to show that fast-moving elephants 'run' with their front legs, but 'walk' with their hind legs.

Although they start this "run" at only 8 km/h, elephants have been reported to reach speeds up to 40 km/h (25 mph), all the while using the same gait. In tests at the Thai Elephant Conservation Centre, the fastest elephants reached a top speed of 18 km/h (11 mph). At this speed, most other four-legged creatures are well into a gallop, even accounting for leg length. Spring-like kinetics could explain the difference between the motion of elephants and other animals.

Ears

Figure: *Difference between Asian (left) and African (right) elephant ears.*

The large flapping ears of an elephant are also very important for temperature regulation. Elephant ears are made of a very thin layer of skin stretched over cartilage and a rich network of blood vessels. On hot days, elephants will flap their ears constantly, creating a slight breeze. This breeze cools the surface blood vessels, and then the cooler blood gets circulated to the rest of the animal's body. The hot blood entering the ears can be cooled as much as 10 °F (6 °C) before returning to the body. Differences in the ear sizes of African and Asian elephants can be explained, in part, by their geographical distribution. Africans originated and stayed near the equator, where it is warmer. Therefore, they have bigger ears. Asians live farther north, in slightly cooler climates, and thus have smaller ears.

The ears are also used in certain displays of aggression and during the males' mating period. If an elephant wants to intimidate a predator or rival, it will spread its ears out wide to make itself look more massive and imposing. During the breeding season, males give off an odor from the musth gland located behind their eyes. Joyce Poole, a well-known elephant researcher, has theorised that the males will fan their ears in an effort to help propel this "elephant cologne" great distances.

Biology and Behaviour

Evolution

The earliest known ancestors of modern-day elephants evolved about 60 million years ago. The ancestor of the elephants from

37 million years ago was aquatic and had a similar lifestyle to a hippopotamus.

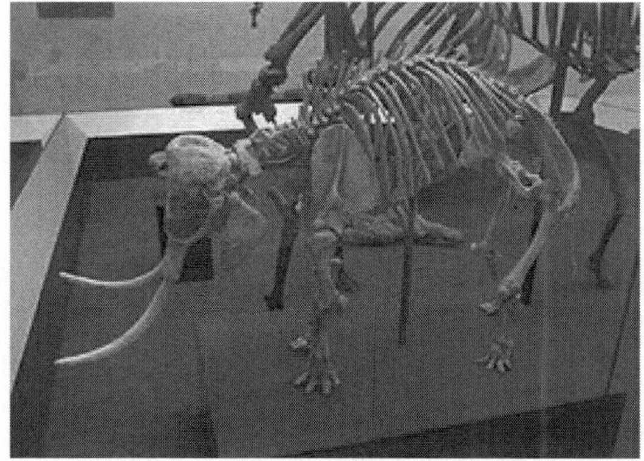

Figure: *The skeleton of a dwarf elephant from the island of Crete. Dwarf elephants were present on some Mediterranean islands until about **10,000** years ago.*

Social Behaviour

Elephants live in a structured social order. The social lives of male and female elephants are very different. The females spend their entire lives in tightly knit family groups made up of mothers, daughters, sisters, and aunts. These groups are led by the eldest female, or matriarch. Adult males, on the other hand, live mostly solitary lives.

The social circle of the female elephant does not end with the small family unit. In addition to encountering the local males that live on the fringes of one or more groups, the female's life also involves interaction with other families, clans, and subpopulations. Most immediate family groups range from five to fifteen adults, as well as a number of immature males and females. When a group gets too big, a few of the elder daughters will break off and form their own small group. They remain very aware of which local herds are relatives and which are not. The life of the adult male is very different.

As he gets older, he begins to spend more time at the edge of the herd, gradually going off on his own for hours or days at a time. Eventually, days become weeks, and somewhere around the age of fourteen, the mature male, or bull, sets out from his natal group for good. While males do live primarily solitary lives, they will occasionally form loose associations with other males. These groups are called bachelor herds. The males spend much more time than the females fighting for dominance with each other. Only the most dominant males will be permitted to breed with cycling females. The less dominant ones must wait their turns. It is usually the older bulls, forty to fifty years old, that do most of the breeding.

The dominance battles between males can look very fierce, but typically they inflict very little injury. Most of the bouts are in the form of aggressive displays and bluffs. Ordinarily, the smaller, younger, and less confident animal will back off before any real damage can be done. However, during the breeding season, the battles can get extremely aggressive, and the occasional elephant is injured. During this season, known as musth, a bull will fight with almost any other male it encounters, and it will spend most of its time hovering around the female herds, trying to find a receptive mate.

In *West with the Night*, Kenyan Aviatrix Beryl Markham suggests the matriarchal society of elephants may be a recent adaptation, since perhaps 1930, to the arrival of firearms. She describes elephant herds containing multiple adult males as well as females. She further describes how the females attempted to hide the males (hunted disproportionately for their tusks) from hunters.

Mating

Figure: *Elephant mating behaviour.*

The mating season is short and females are only able to conceive for a few days each year. She will detach herself from the herd. The scent of the female (cow) elephant in heat (or estrus) attracts the male and she also uses audible signals to attract the male. As the female can usually outrun the male, she does not have to mate with every male that approaches her. The male initiates the courtship and the female ignores him for several minutes. He then stops and starts again. Elephants display a range of affectionate interactions, such as nuzzling, trunk intertwining, and placing their trunks in each other's mouths. In a rarely observed display of his affection, he may drape his trunk outside of his tusks during the ritual. The interactions may last for

20–30 minutes and do not necessarily result in the male mounting the female, though he may demonstrate arousal during the ritual. The female elephant is not passive in the ritual and uses the same techniques as the male.

African as well as Asiatic males will engage in same-sex bonding and mounting. The encounters are analogous to heterosexual bouts, one male often extending his trunk along the other's back and pushing forward with his tusks to signify his intention to mount. Unlike heterosexual relations, which are always of a fleeting nature, those between males result in a "companionship", consisting of an older individual and one or two younger, attendant males. Same-sex relations are common and frequent in both sexes, with Asiatic elephants in captivity devoting roughly 46% of sexual encounters to same-sex activity.

Elephant Cognition

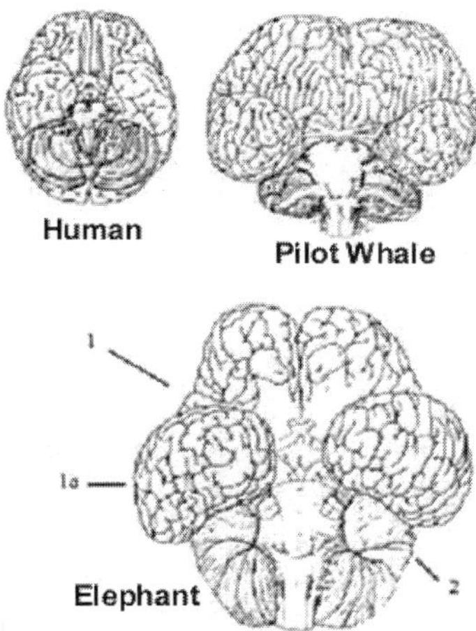

Figure: *Human, pilot whale and elephant brains up to scale. (1)-cerebrum (1a)-temporal lobe and (2)-cerebellum*

Elephants are amongst the world's most intelligent species. With a mass of just over 5 kg (11 lb), elephant brains are larger than those of any other land animal, and although the largest whales have body masses twenty-fold those of a typical elephant, whale brains are barely twice the mass of an elephant's brain. The elephant's brain is similar to that of humans in terms of structure and complexity - such as the elephant's cortex having as many neurons as a human brain, suggesting convergent evolution.

Elephants exhibit a wide variety of behaviours, including those associated with grief, learning, allomothering, mimicry, art, play, a sense of humor, altruism, use of tools, compassion, cooperation, self-awareness, memory and possibly language. All indicate that Elephants are highly intelligent; it is thought they are equal with cetaceans and primates in this regard. Due to the high intelligence and strong family ties of elephants, some researchers argue it is morally wrong for humans to cull them.

Aristotle once said that elephants were "The beast which passeth all others in wit and mind."

Brain Structure

Cerebral Cortex

The elephant (both Asian and African) has a very large and highly convoluted neocortex, a trait also shared by humans, apes and certain dolphin species. Scientists see this as a sign of complex intelligence. While this is the widely held belief, there is at least one exception: the echidna has a highly developed brain, yet is not considered to be intelligent.

Asian elephants have the greatest volume of cerebral cortex available for cognitive processing of all existing land animals. Elephants have a volume of cerebral cortex available for cognitive processing that exceeds that of any primate species, and extensive studies place elephants in the category of great apes in terms of cognitive abilities for tool use and tool making.

The elephant brain exhibits a gyral pattern more complex and with more numerous convolutes, or brain folds, than that of humans, primates or carnivores, but less complex than cetaceans. However, the cortex of the elephant brain is "thicker than that of cetaceans" and is believed to have as many cortical neurons (nerve cells) and cortical synapses as that of humans, which exceeds that of cetaceans. Elephants are believed to rank equal with dolphins in terms of problem-solving abilities, and many scientists tend to rank elephant intelligence at the same level as cetaceans; in fact, a 2011 article published by ABC Science states that, "elephants [are as] smart as chimps, [and] dolphins".

Other Features of the Brain

Elephants also have a very large and highly convoluted hippocampus, a brain structure in the limbic system that is much bigger than that of any human, primate or cetacean. The hippocampus of an elephant takes up about 0.7% of the central structures of the brain, comparable to 0.5% for humans and with 0.1% in Risso's dolphins and 0.05% in bottlenose dolphins.

The hippocampus is linked to emotion through the processing of certain types of memory, especially spacial. This is thought to be possibly why elephants suffer from psychological flashbacks and the equivalent of post-traumatic stress disorder (PTSD).

The encephalisation quotient (EQ) of elephants ranges from 1.13 to 2.36. The average EQ is 2.14 for Asian elephants, and 1.67 for African, with the overall average being 1.88. There is considerable debate about whether the EQ, or size of the brain relative to body size, is an accurate gauge for intelligence, especially since the animal with the highest EQ is a treeshrew which are not considered to be highly intelligent.

Brain Size at Birth Relative to Adult Brain Size

Like humans, elephants must learn behaviour as they grow up. They are not born with the instincts of how to survive. Elephants have a very long period in their lives for learning, lasting for around ten years. One comparative way to try to gauge intelligence is to compare brain size at birth to the fully developed adult brain. This indicates how much learning a species accumulates while young. The majority of mammals are born with a brain close to 90% of the adult weight, while Humans are born with 28%, bottlenose dolphins with 42.5%, chimpanzees with 54%, and elephants with 35%. This indicates that elephants have the highest amount of learning to undergo next to humans, and behaviour is not mere instinct but must be taught throughout life. It should be noted that instinct is quite different from learned intelligence. Parents will teach their young how to feed, use tools and learn their place in the highly complex elephant society. The cerebrum temporal lobes, which function as storage of memory, are much larger than those of a human.

Spindle Neurons

Spindle cells appear to play a central role in the development of intelligent behaviour. Initially, it was thought that the presence of spindle neurons was unique to humans and the great apes. However, studies have shown that spindle neurons are also found in the brains of both Asian and African elephants, as well as humpback whales, fin whales, killer whales, sperm whales, bottlenose dolphins, Risso's dolphins, and beluga whales. The remarkable similarity between the elephant brain and the human brain supports the thesis of convergent evolution.

Elephant Society

The elephant has one of the most closely knit societies of any living species. Elephant families can only be separated by death or capture.

Cynthia Moss, an ethologist specialising in elephants, recalls an event involving a family of African elephants. Two members of the family were shot by poachers, who were subsequently chased off by the remaining elephants. Although one of the elephants died, the other, named Tina, remained standing, but with knees beginning to give way. Two family members, Trista and Teresia (Tina's mother), walked to both sides of Tina and leaned in to hold her up. Eventually, Tina grew so weak, she fell to the ground and died.

However, Trista and Teresia did not give up but continually tried to lift her. They managed to get Tina into a sitting position, but her body was lifeless and fell to the ground again. As the other elephant family members became more intensely involved in the aid, they tried to put grass into Tina's mouth. Teresia then put her tusks beneath Tina's head and front quarters and proceeded to lift her. As she did so, her right tusk broke completely off, right up to the lip and nerve cavity. The elephants gave up trying to lift Tina but did not leave her; instead, they began to bury her in a shallow grave and throw leaves over her body. They stood over Tina for the night and then began to leave in the morning. The last to leave was Teresia.

Because elephants are so closely knit and highly matriarchal, a family can be devastated by the death of another (especially a matriarch), and some groups never recover their organisation. Cynthia Moss has observed a mother, after the death of her calf, walk sluggishly at the back of a family for many days.

Edward Topsell stated in his publication *The History of Four-Footed Beasts* in 1658, "There is no creature among all the Beasts of the world which hath so great and ample demonstration of the power and wisdom of almighty God as the elephant." Of note, he stated in the same publication that elephants worship the sun and the moon and become pregnant by chewing on mandrake, neither of which is true. Elephants are believed to be on par with chimpanzees with regards to their cooperative skills.

Elephant Altruism

Elephants are thought to be highly altruistic animals that will even aid other species, including humans, in distress. In India, an elephant was helping locals lift logs by following a truck and placing the logs in pre-dug holes upon instruction from the mahout (elephant trainer). At a certain hole, the elephant refused to lower the log. The mahout came to investigate the hold-up and noticed a dog sleeping in the hole. The elephant only lowered the log when the dog was gone.

Cynthia Moss has often seen elephants going out of their way to avoid hurting or killing a human, even when it was difficult for them (such as having to walk backwards to avoid a person).

Joyce Poole documented an encounter told to her by Colin Francombe on Kuki Gallman's Laikipia Ranch. A ranch herder was out on his own with camels when he came across a family of elephants. The matriarch charged at him and knocked him over with her trunk, breaking one of his legs. In the evening, when he did not return, a search party was sent in a truck to find him. When the party discovered him, he was being guarded by an elephant. The animal charged the truck, so they shot over her and scared her away. The herdsman later told them that when he could not stand up, the elephant used her trunk to lift him under the shade of a tree. She guarded him for the day and would gently touch him with her trunk.

Self-medication

Elephants in Africa will self-medicate by chewing on the leaves of a tree from the Boraginaceae family, which induces labor. Kenyans also use this tree for the same purpose.

Death Ritual

Elephants are the only species on Earth other than *Homo sapiens sapiens* and Neanderthals known to have or have had any recognisable ritual around death. They show a keen interest in the bones of their own kind (even unrelated elephants that have died long ago). They are often seen gently investigating the bones with their trunks and feet and remaining very quiet. Sometimes elephants that are completely unrelated to the deceased will still visit their graves. When an elephant is hurt, other elephants (even if they are unrelated) will aid them.

Elephant researcher Martin Meredith recalls an occurrence in his book about a typical elephant death ritual that was witnessed by Anthony Hall-Martin, a South African biologist who had studied elephants in Addo, South Africa, for over eight years. The entire family of a dead matriarch, including her young calf, were all gently touching her body with their trunks, trying to lift her. The elephant herd were all rumbling loudly. The calf was observed to be weeping and made sounds that sounded like a scream, but then the entire herd fell incredibly silent. They then began to throw leaves and dirt over the body and broke off tree branches to cover her. They spent the next two days quietly standing over her body. They sometimes had to leave to get water or food, but they would always return.

Occurrences of elephants behaving this way around human beings are common throughout Africa. On many occasions, they have buried dead or sleeping humans or aided them when they were hurt. Meredith also recalls an event told to him by George Adamson, a Kenyan Game Warden, regarding an old Turkana woman who fell asleep under a tree after losing her way home. When she woke up, there was an elephant standing over her, gently touching her. She kept very still, because she was very frightened. As other elephants arrived, they began to scream loudly and buried her under branches. She was found the next morning by the local herdsmen, unharmed.

George Adamson also recalls when he shot a bull elephant from a herd that kept breaking into the government gardens of Northern Kenya. George gave the elephant's meat to local Turkana tribesmen and then dragged the rest of the carcass half a mile away. That night, the other elephants found the body and took the shoulder blade and leg bone and returned the bones to the exact spot the elephant was killed. Scientists often debate the extent that elephants feel emotion.

Play

Joyce Poole on many occasions has observed wild African elephants at play. They apparently do things for their own and others' entertainment. Elephants have been seen sucking up water, holding their trunk high in the air, and then spraying the water like a fountain.

Mimicry

Recent studies have shown that elephants can also mimic sounds they hear. The discovery was found when Mlaika, an orphaned elephant, would copy the sound of trucks passing by. So far, the only other animals that are thought to mimic sounds are whales, dolphins, bats, primates, clouded leopards, and birds. Calimero, an African elephant who was 23 years old, also exhibited a unique form of mimicry. He was in a Swiss zoo with some Asian elephants. Asian elephants use chirps that are different from African elephants' deep rumbling noises. Calimero also began to chirp and not make the deep calls that his species normally would.

Kosik, an Indian elephant at Everland Amusement Park, South Korea, surprised trainers when they thought there was a person in his enclosure that was actually Kosik imitating Jong Gap Kim, his trainer. Kosik can make sounds imitating up to eight Korean words, including *sit, no, yes* and *lie down*. His mimicry is remarkably human-sounding. Kosik produces human-like sounds by putting his trunk in his mouth and then shaking it while breathing out, similar to how people whistle with their fingers. Elephants use contact calls to stay in touch with one another when they are out of one another's sight. Female elephants are able to remember and distinguish the contact calls of female family and bond group members from those of females outside of their extended family network. They can also distinguish between the calls of family units depending upon how frequently they came across them.

Tool Use

Elephants show a remarkable ability to use tools, using their trunks like arms. Elephants have been observed digging holes to drink water and then ripping bark from a tree, chewing it into the shape of a ball, filling in the hole and covering over it with sand to avoid evaporation, then later going back to the spot for a drink. They also often use branches to swat flies or scratch themselves. Elephants have also been known to drop very large rocks onto an electric fence either to ruin the fence or to cut off the electricity.

Art

Like several other species, elephants are able to produce abstract art using their trunks to hold brushes. An example of this was shown in the TV program *Extraordinary Animals*, in which elephants at a camp in Thailand were able to draw self-portraits with flowers. Although the images were drawn by the elephants, there was always a person assisting and guiding the movement. From those presentations, it cannot be definitely evaluated whether the elephants are conscious about the shape of their drawings or not.

This extraordinary video documentation of an elephant painting a picture of an elephant—possibly indicating self-awareness—has become widespread on internet news and video websites. The quality of the painting is extremely high, leading many astonished viewers to doubt the video's authenticity. The website snopes.com, which specialises in debunking urban legends, lists the video as "true", in that the elephant produced the brush strokes, but notes that the similarity of the produced paintings is indicative of a learned sequence of strokes rather than a creative effort on the part of the elephant.

Problem-solving Ability

Elephants are able to spend a lot of time working on problems. They are able to change their behaviour radically to face new challenges, a hallmark of complex intelligence. A 2010 experiment revealed that in order to reach food, "elephants can learn to coordinate with a partner in a task requiring two individuals to simultaneously pull two ends of the same rope to obtain a reward", putting them on an equal footing with chimpanzees in terms of their level of cooperative skills.

In the 1970s, at Marine World Africa, USA, there lived an Asian elephant named Bandula. Bandula worked out how to break open or unlock several of the pieces of equipment used to keep the shackles on her feet secure. The most complex device was a brommel hook, a device that closes when two opposite points are slid together. Bandula used to fiddle with the hook until it slid apart when it was aligned. Once she had freed herself, she would help the other elephants escape. In Bandula's case and certainly with other captive elephants, there was an element of deception involved during the escapes, such as the animals looking around making sure no one was watching.

In another case, a female elephant worked out how she could unscrew iron rods with an eye hole that were an inch thick. She used her trunk to create leverage and then untwist the bolt.

Ruby, an Asian elephant at Phoenix Zoo would often eavesdrop on conversations keepers would have talking about her. When she heard the word *paint*, she became very excitable. The colours she favoured were green, yellow, blue and red. On one particular day, there was a fire truck that came and parked outside her enclosure where a man had just had a heart attack. The lights on the truck were flashing red, white and yellow. When Ruby painted later on in the day, she chose those colours. She also showed a preference for particular colours that the keepers wore.

Harry Peachey, an elephant trainer, developed a cooperative relationship with an elephant named Koko. Koko would help out the keepers, "prompting" the keepers to encourage him with various commands and words that Koko would learn. Peachey stated that elephants are almost predisposed to cooperate and work with humans as long as they are treated with respect and sensitivity. Koko worked out when his keepers needed a bit of "elephant help" when they were transferring the females of the group to another zoo. When the keepers wanted to transfer a female, they would usually say her name, followed by the word *transfer* (e.g., "Connie transfer"). Koko soon figured out what this meant. If the keepers asked an elephant to transfer and they did not budge, they would say, "Koko, give me a hand." When he heard this, Koko would help. Peachey firmly believes that after 27 years of working with elephants, they can understand the semantics and syntax of some of the words they hear. This is something thought to be very rare in the animal kingdom.

A study by Dr. Naoko Irie of Tokyo University has shown that elephants demonstrate skills at arithmetic. The experiment "consist[ed] of dropping varying numbers of apples into two buckets in front of the [Ueno Zoo] elephants and then recording how often they could correctly choose the bucket holding the most fruit." When more than one apple was being dropped into the bucket, this meant that the elephants had to "keep running totals in their heads to keep track of the count." The results showed that "Seventy-four percent of the time, the animals correctly picked the fullest bucket. An African elephant named Ashya scored the highest with an amazing eighty-seven percent ... Humans in this same contest managed a success rate of just sixty-seven percent." The study was also filmed to ensure its accuracy.

Self-awareness

Asian Elephants have joined a small group of animals, including great apes, bottlenose dolphins and magpies, that exhibit self-awareness. The study was conducted with the Wildlife Conservation Society (WCS) using elephants at the Bronx Zoo in New York. Although many animals will respond to a mirror, very few show any evidence that they recognise it is in fact themselves in the mirror reflection.

The Asian elephants in the study also displayed this type of behaviour when standing in front of a 2.5 m-by-2.5 m mirror - they inspected the rear and brought food close to the mirror for consumption.

Evidence of elephant self awareness was shown when the elephant Happy repeatedly touched a painted *X* on her head with her trunk, a mark which could only be seen in the mirror. Happy ignored another mark made with colourless paint that was also on her forehead to ensure she was not merely reacting to a smell or feeling.

Frans De Waal, who ran the study, stated, "These parallels between humans and elephants suggest a convergent cognitive evolution possibly related to complex society and cooperation."

Joyce Poole, of the Amboseli Elephant Research Project, Kenya, has demonstrated vocal learning and imitation in elephants of sounds made by each other and in the environment. She is beginning to research whether sounds made by elephants have dialects, a trait that is rare in the animal kingdom.

Self-awareness and Culling

There has been considerable debate over the issue of culling African elephants in South Africa's Kruger National Park as a means of controlling the population. Some scientists and environmentalists argue that it is "unnecessary and inhumane" to cull them since "elephants resemble humans in a number of ways, not least by having massive brains, social bonds that appear to be empathetic, long gestations, high intelligence, offspring that require an extended period of dependent care, and long life spans." A South African Animal Rights group asked in a statement anticipating the announcement, "How much like us do elephants have to be before killing them becomes murder?". Others argue that culling is necessary when biodiversity is threatened.

Elephant Digestive System

The digestive system of the elephant is set apart almost exclusively by its sheer size. Apart from this factor, it is very similar to the digestive systems of most other mammals. Likewise, the liver and the pancreas, both key organs that aid in digestion (albeit indirectly), are also much the same as ours, only far larger in scale.

The mouth, where food begins its digestive process, is relatively small compared to the enormous animal itself. In addition, this mouth is not able to open very wide either. Digestion in the mouth is aided by the ample saliva produced by the many salivary glands in the mouth as well as by the mucus glands in the oesophagus. Saliva contains enzymes that begin breaking the food down. It also serves along with the mucus to lubricate the tough vegetation eaten by the elephants, which even includes thorns and bark. The muscular oesophagus ushers the food into the stomach, which acts more as a storage sac as not much digestion takes place in this organ. The elephant's stomach is cylindrical in shape and the middle of the organ is particularly glandular. From here, food is ushered into the extraordinarily long intestines. The intestine of an elephant can reach up to 19 metres in length!

The intestine is where most of the digestion of the vegetative diet takes place. At the point at which the small intestine meets the large one (or colon), bacteria aids in the fermentative digestion of the cellulose (typical of this diet). This location is called the caecum and is particularly rich in blood vessels. The caecum is divided into many smaller sacs and the products of digestion are absorbed through its relatively thin walls.

Because the elephant only digests and makes good use of 40% of its intake, the intestine is also instrumental in the formation or faeces and the efficient absorption of water. The size of the faeces is often used to determine the age of the elephant as it retains the shape formed by the walls of the rectum, indicating its size.

Because the elephant absorbs so few nutrients from the food it ingests, the dung is rich in nutrients and solid food matter. Therefore, it is beneficial to many other animals, as they are able to feed off of the relatively untouched constituents of the faeces. Common beneficiaries include dung beetles and many species of bird.

An elephant will spend much of the day searching for food to ensure that, although they actually digest relatively little from their food, they are taking in enough to remain healthy. In areas where vegetation is plentiful, the elephant will spend approximately 16 hours a day grazing or eating fruits, nuts and seeds from trees. In areas that are very dry or lacking in healthy plants and grass, the animal may need to spend up to 24 hours a day in search of food. For correct nutrition, the elephant needs to consume between 140kg and 270kg of food every day.

Elephant Altruism

Altruism may be defined as a selfless concern for the welfare of others. Elephants are particularly altruistic. However, their altruism extends not only to other elephants, but to many other species in distress as well, including humans.

Elephants truly demonstrate their incredible insight and ability to read the needs and desires of other living creatures. Not only do they perceive these; they also act in such a way as to accommodate certain needs and even satisfy these in some cases. At times, elephants will assist another creature at the risk of its own safety. This capacity to act based on compassion and not survival instincts is even rare in human beings.

There are many examples the world over of elephants who have displayed this advanced sense of altruism with no benefit to themselves. One is that of a cow who relentlessly worked to save a baby rhinoceros that was stuck in the mud despite the attacks of its mother on the elephant. Another astounding record tells of a ranch herder who was struck off his camel by a charging matriarch. His leg was broken and he was alone. When the search party was dispatched and eventually found him that evening, he had been dragged to a shaded area and protected all the while by one of the other cows, who had split from the herd to return to him and regularly touched him with her trunk in a gesture of comfort and reassurance. A working elephant in India refused to lower one of the logs it was carrying that day. When the elephant trainer walked in front of her to see why she was not cooperating, he found a dog sleeping in the hole in which she was to drop her log. She would not lower the timber until the dog had moved out of danger.

This sense of altruism is also clear amongst the members of the herd. The birth of a calf is a herd activity, as each cow participates and celebrates in the event. Even after its birth, the calf is cared for by the other cows; being allowed to suckle from any female able to provide it with milk. Not only does this ensure that the herd is as strong and healthy as possible in a physical sense, but it also strengthens the individual bonds and reinforces the sense of altruism.

The matriarch, which is usually the strongest, largest and oldest female of the herd, has the responsibility to protect her herd. In true altruistic style, she will defend and protect them at any cost, even at that of her own valuable life.

Despite the sheer horror inflicted on such magnificent animals by human beings, elephants have even been known to either avoid harming a human, or even protecting them, as shown by the example above. In some cases, the elephant has had to retreat from the situation itself in order to ensure that the human is not afraid or harmed. Considering the elephant's memory and insight, it is remarkable that it would choose not to harm the beast that has contributed so much to its slaughter and mistreatment. This is further testimony to the intellectual and social superiority of the beautiful elephant.

Elephant Emotion – Joy

Elephants have always cultivated a fascination for human beings, a fact that is evident by the many myths and folklore tales about them. Researchers have observed these animals at length, and they never seem to disappoint. In fact, despite much data being gleaned on elephants, they still have the ability to surprise their trainers with their unique abilities and unmatched intelligence. One of the qualities that set elephants apart as animals (and evoke a sense of kinship to human beings) is that of emotion. Elephants are capable of sadness, joy, love, jealousy, fury, grief, compassion and distress.

Elephants are one of a select few animals that have the capacity to be joyful and playful with one another, to grasp humour and appreciate it. As social creatures, elephants will frequently touch one another in affectionate, loving ways. Joy is most often displayed when they greet close friends or family members. Herds sometimes split and larger families are separated, depending on the matriarch's decision. This can be due to excessive numbers or shortages of food or water. When these herds meet at watering holes or breeding spots, they joyfully greet one another. This welcoming reception includes turning around in circles, holding their heads up, flapping their ears, trumpeting, screaming and even urinating and defecating. Elephants who have formed very close bonds with people are also likely to react in this way on seeing their companion after a separation.

Another major cause for celebration is the birth of a calf. During the birth, the aunts and matriarch gather around the mother in joyful support. Once the calf is born, the others help it to its mother's teat to begin suckling. Celebrations begin and the cows begin to trumpet, rumble and even scream in joy and excitement.

Interestingly, elephants have even been known to amuse themselves by playing games. These can either be played in a group, or with just one player. The animal will use objects from its environment and toss, twist, or interact with it in some way. Games are initiated by trumpeting loudly, indicating to those in the herd that a new session has begun. Unlike humans and some other species, playing games is not confined to the youth; older matriarchs and bulls have been known to engage in some playful recreation. Elephants have even displayed a sense of humour in their games, often tricking and teasing their spectators. Elephants in captivity have been observed stealing from or playfully handling the onlookers.

It is doubtful that we will ever have a full understanding of the capacity of the elephant brain. As we submit humbly to this fact, we continue to enjoy the surprises as we observe them unfolding.

Elephant Emotions - Grieving

Elephants evoke a sense of fascination in human beings and have done so for centuries. Research and observation has yielded many intriguing facts about these animals. Perhaps the most enticing quality of the elephant is its undeniable similarity to us, manifested by the close bonds they form with family members, their communication, life span, the care of their young and their emotions. Elephants experience many of the same emotions as people do, ones that are usually restricted to being that of humans', seldom seen in animals. Elephants are capable of sadness, joy, love, jealousy, fury, grief, compassion and distress.

The elephant's capacity for sadness and grief is truly unique amongst members of the animal world, as it is particularly complex in terms of emotions. While most animals do not hesitate to leave the weak and young behind to die, elephants are distressed by the situation, and continue to show signs of this grieving for extended periods of time.

Because elephants live in such close-knit herds and live for about as long as humans do (approximately 70 years), they form strong bonds with those around them. When these ones die, the rest of the herd mourns that death. Mothers and aunts are also prone to mourning a still-born calf. The mother of a dead calf (whether at birth or later on in its life) shows her grief through her physical disposition. Her eyes are sunken and her ears drooping, her mood is visibly miserable.

The herd will take great care in the burial of the dead. Cows walk to and fro in search of leaves and twigs. They use this to cover the body of the deceased in an act of dignity for the dead. When a herd encounters the skeleton of a dead elephant, they have shown an undeniable fascination with the bones. T

he cows will mull over the bones, fondling them in thoughtful contemplation. Cows take bones from the skeleton and scatter them, hiding them under bushes in the surrounding area. This behaviour is thought to be as a protection for the rest of the herd, as it throws stalking predators off the trail of the cows and their calves. Even years later, elephant have been observed revisiting the site where one of their herd or family had died. They will remain here for days at a time, mourning the loss of that one.

It is when we consider this level of grief and love for one another that we can truly admire the role of the matriarch. When a weak or aged elephant is jeopardising the safety of the herd, the wise old leader has to make the very difficult decision to leave that one behind to protect the rest of her family. In obedience and out of respect for her authority, the other females will follow her guidance, submitting willingly. Their grief must be overwhelming, but their loyalty to the matriarch is even stronger.

Elephant Evolution

The mighty elephant is both beautiful and functional in design. Assuming that this level of development was as a result of evolution (versus creation), it is thought that the first elephant ancestors began their trek through the evolutionary cycle 60 million years ago. It was at this time that Condylartha (herbivorous mammals) evolved into Ungulata (the Latin word for "provided with hooves"). From here, according to Eisenberg, the ungulate split into 5 groups:

1. Eparctocyon (animals with cloven hooves or even toes)
2. Cete (whales and dolphins)
3. Phenacodonta (horses, tapirs and rhinos)
4. Meridungulata (an extinct species)
5. Paenungulata (superorder – elephant ancestors).

It is thought that the modern-day elephant evolved from an aquatic animal that used its trunk as a sort of snorkel while it spent extended periods of time beneath the surface of the water. Even today, elephants are able to swim in this way, staying underwater for hours at a time, with only the tip of their trunk exposed. According to these theories, elephants (Proboscidea) are closely related to manatees and dugong (Sirenia). It is deduced, then, that Proboscidea and Sirenia must share a common ancestor, from which each developed into different species. The Paenungulata group has been identified as the elephant ancestor by intensive DNA research, which revealed that the hyraxes branch off earlier, while the Proboscidea and Sirenia are still closely related in Tethytheria.

"Proboscidea" is a Greek word that means "having a nose". Over 350 Proboscidea have been identified as having existed over the last 50 million years. The only areas that were not found to have been home at least some of these species are Antarctica and Australia. All but 3 of these species are now extinct, leaving only the Asian Elephant, African Savanna Elephant (the world's largest land mammal) and African Forest Elephant. Many of the more delicate and specialised species are likely to have died due to environmental changes that, when coupled with slow reproductive rates, spelt extinction.

The African species can be found in nearly 40 African countries. They are larger than their Asian counterparts. They are further distinguished by different ear- and head shapes. Originally, the two different African elephants were considered to be subspecies. However, DNA research has revealed that they are actually two separate species. This classification has important ramifications in terms of conservation and the population statistics.

Research continues in an effort to bridge the gaps that exist in the evolutionary development of these extraordinary animals. There remain mysteries regarding associated species and the exact path these ancestors followed before evolving into the elephant as we know it today.

Social Order of Elephants

Elephants are known for their superior intelligence as well as their structured social order. One of the main characteristics of the social order in the herd is that males and females live entirely different and separate lives.

Elephants are a matriarchal society; that is, one that is led by a head cow, who presides over her herd of females. Each herd is made up of mothers, daughters, sisters and aunts. They are guided by the oldest and largest female of the herd. This herd sticks closely together, rejoicing at the birth of a calf and mourning at the death of a member.

The herd of females, although maintaining close bonds among themselves, also interacts well with other herds, families, and clans. An average herd of immediate family will comprise of 5 to 15 adult elephants as well as immature males and females. As the herd grows, some members split to form new herds. In this way, families are divided and allowed to expand outwards. However, these members never forget their family roots and commit much time and effort to keeping track of their relatives through vocal and non-vocal communication.

The male, on the other hand, lives apart from the matriarchal herd, and travels alone or with other males in a bachelor pod. The drift from the herd starts during adolescence, at which time the young bulls start to spend less and less time with the herd. Eventually, the break is made completely. After this distance is established, the bulls will live solitary lives, mingling with the females of the family only for mating purposes. The bonds between the males in a bachelor pod are loosely established and loyalty is fickle. Males will frequently fight with other bulls over dominance as only the strongest are able to mate with receptive females. This ensures strong, healthy calves. When males fight over dominance, there are usually few injuries, although the fighting seems quite violent to onlookers. It is only during the intense breeding season that these encounters become somewhat more intense.

Same-sex bonding is common in elephants. This is usually in the form of affectionate trunk-entwining, kissing and placing trunks into the other animal's mouth. Males will actually mount one another. Unlike the fleeting encounter between a bull and a cow, this mounting will last for much longer and often consists of an older adult with one or two juvenile males. When one male splits from all other company and becomes excessively aggressive and violent, he is known as the rogue male. He will usually win at any domination battles and, therefore, mate with more females than the other bulls. This means that his offspring are strong and that the herd enjoys fairly pure genes. Because elephants enjoy such rigid social structures and norms, they are able to form close bonds with those within their immediate herd and family, with females benefitting particularly from this security. This allows an enormous degree of trust and sanctuary within the herd.

Chapter 4

Behaviour of Fish

Aquatic Predation

Aquatic predation presents a special difficulty as compared to predation on land, because the density of water is about the same as that of the prey, so that the prey tends to be pushed away. This problem was first identified by R.N. Alexander. As a result, underwater predators, especially bony fish, have evolved a number of specialised feeding mechanisms, known as filter feeding, ram feeding, suction feeding, protrusion, and pivot feeding.

Most underwater predators combine more than one of these basic principles. For example a typical generalised predator, such as the cod, combines suction with some amount of protrusion and pivot feeding.

Suction Feeding

Suction feeding is a method of ingesting a prey item in fluids by sucking the prey into the predator's mouth. This is typically accomplished by the predator expanding the volume of its oral cavity and/or throat, resulting in a pressure difference between the inside of the mouth and the outside environment. When the mouth is opened, the pressure difference causes water to flow into the predator's mouth, carrying the prey item in with the fluid flow. Bony fish have evolved ingenious mechanisms, so-called mechanical linkages for rapid and forceful expansion of the buccal cavity.

Ram Feeding

Ram feeding, also known as lunge feeding, is a method of feeding underwater in which the predator moves forward with its mouth open, engulfing the prey along with the water surrounding it. During ram feeding, the prey remains fixed in space, and the predator moves its jaws past the prey to capture it. The motion of the head may induce a bow wave in the fluid which pushes the prey away from the jaws, but this can be avoided by allowing water to flow through the jaw. This can be accomplished by means of an expandable throat, as in snapping turtles and baleen whales, or by allowing water to flow out through the gills, as in sharks and herring. A number of species have evolved narrow snouts, as in gar fish and water snakes .

Herrings often hunt copepods. If they encounter copepods schooling in high concentrations, the herrings switch to ram feeding. They swim with their mouth wide open and their opercula fully expanded. Every several feet, they close and clean their gill rakers for a few milliseconds (filter feeding). The fish all open their mouths and opercula wide at the same time. The fish swim in a grid where the distance between them is the same as the jump length of the copepods.

Pivot Feeding

Pivot feeding is a method to transport the mouth towards the prey by an upward turning of the head, which is pivoting on the neck joint. Pipefish such as sea horses and sea dragons are specialised on this feeding mechanism. With prey capture times of down to 5 ms (shrimpfish *Centriscus scutatus*) this method is used by the fastest feeders in the animal kingdom.The secret of the speed of pivot feeding is in a locking mechanism, in which the hyoid arch is folded under the head and is aligned with the urohyal which connects to the shoulder girdle. The trigger mechanism of unlocking is debated, but is probably in lateral adduction.

Protrusion

Protrusion is the extension of the mouth towards the prey, via mechanical linkages. Protrusion is only known in modern bony fishes, which possess many forms of coupled linkages in their head. Remarkable examples are the slingjaw wrasse and the sand eel which can protrude their mouth by several centimetres.

Filter and Suspension Feeding

This is the selection of food particles from a water flow, for example by the gill rakers of fish, the baleens of whales, or the ostia of sponges.

Filter Feeding

In filter feeding, the water flow is primarily generated by the organism itself, for example by creating a pressure, by active swimming, or by ciliary movements.

Suspension Feeding

In suspension feeding, the water flow is primarily external or if the particles themselves move with respect to the ambient water, such as in sea lilies.

Bait Ball

A bait ball, or baitball, occurs when small fish swarm in a tightly packed spherical formation about a common centre. It is a last ditch defensive measure adopted by small schooling fish when they are threatened by predators. Small schooling fish are eaten by many types of predators, and for this reason they are called bait fish or forage fish.

For example, sardines group together when they are threatened. This instinctual behaviour is a defence mechanism, as lone individuals are more likely to be eaten than large groups. Sardine bait balls can be 10–20 metres in diametre and extend to a depth of 10 metres. The bait balls are short lived and seldom last longer than 10 minutes.

However, bait balls are also conspicuous, and when schooling fish form a bait ball they can draw the attention of many other predators. As a response to the defensive capabilities of schooling fish, some predators have developed sophisticated countermeasures. These countermeasures can be spectacularly successful, and can seriously undermine the defensive value of forming bait balls.

Background

Small pelagic fish live in the open water, so unlike demersal or reef fish, they cannot hide among kelp, or in crevices in coral, or under rocks on the bottom. This leaves them vulnerable to attack by large predatory fish, as well as other predators, such as marine mammals and seabirds. As a result, small pelagic fish usually aggregate together in schools for protection. Schooling fish have evolved sophisticated evasion techniques. When they school they have many eyes, which makes ambush difficult, and their silvery bodies dazzle, which makes it difficult for predators to pick out individual fish. They react to movements from a predator with lightning reflexes, rhythmically streaming up and down with rapid direction changes. When a predator approaches, they can split and reform behind the predator. During the final stages of an attack, they can explosively disband in all directions and then just as rapidly reform. Defensive manoeuvres like these appear to be choreographed, though they are not. Within the school itself, there is no centralised intelligence aware of how the school has configured itself. Rather, the schooling behaviour is the emergent consequence of relatively simple rules followed by each individual fish in the school, such as remaining close together, moving in the same direction, and avoiding collisions with each other.

Some species of forage fish, driven by nutrient availability and their life-cycle stage, form vast schools at predictable locations and times of the year. Normally, schooling works well as protection from occasional predators. But when fish school in vast numbers they can attract correspondingly huge numbers of predators, including seabirds, sharks, tuna, billfish, pods of dolphins, and killer and humpback whales. Aggregations of predators on such a scale means the schools can be attacked on all sides, and panicked into forming bait balls.

A bait ball is a last ditch defensive measure adopted by fish schools when they are overwhelmed and more effective defensive strategies have broken down. Schooling fish are easier to attack once they abandon their free streaming behaviour and form into a tight bait ball. Many predator species have learned that by interacting cooperatively they can panic schooling fish into forming a bait ball. This cooperative behaviour can occur both intraspecifically (among the individuals within a predator species) and interspecifically (across individuals belonging to more than one predator species).

Formation and Dissolution

The process that leads to the formation of a bait ball typically starts when predators locate a fish school deep below the surface. The predators make rushes and use various scare tactics to force the fish school to the surface, herding it at the same time into a compact volume. The alarmed fish, trapped against the surface above and surrounded all about, abandon their coordinated schooling movements and become chaotic. Their graceful and disciplined schooling strategies of uniform spacing and polarity degrade into frenetic attempts by each fish to save itself. In this way, a dense bait ball forms as each fish scrambles to get away from the surface of the ball and hide in the interior. The symmetry of this centripetal action forms a sphere, the shape with the minimum surface area for a given volume. This exposes the fewest number of fish on the surface to the predators. The movement, sound and smell can attract more predators, including different predator species, until there is a carousel of them, each species using their own characteristic predatory strategies. Fish that break loose are singled out and eaten. A frenzy can develop as predators compete, the water reddening with blood as shredded flesh and scales drift to the depths. As the bait ball reduces in size and number, it becomes progressively easier for the predators to target the remaining survivors.

Predator Strategies

Predators have devised various countermeasures to disrupt the defensive shoaling and schooling manoeuvres of forage fish. Often this involves charging the school or bait ball at high speed. Some whales *lunge feed* on bait balls. Lunge feeding is an extreme feeding method, in which the whale accelerates from below a bait ball to a high velocity and then opens its mouth to a large gape angle. This generates the water pressure required to expand its mouth and engulf and filter a huge amount of water and fish. Lunge feeding by the huge rorqual whales is said to be the largest biomechanical event on Earth.

Swordfish charge at high speeds through forage fish schools, slashing with their swords to kill or stun prey. They then turn and return to consume their catch. Thresher sharks use their long tails to stun shoaling fishes. Spinner sharks charge vertically through schools, spinning on their axis with their mouths open and snapping all around. The shark's momentum at the end of these spiralling runs often carries it into the air.

Gannets plummet from heights of 30 metres (100 feet), plunging through the water leaving vapour-like trails behind like fighter planes. They enter the water at speeds up to 86 kilometres per hour (53 mph) and descend to depths of 34 metres (111 feet). Gannets have air sacs under their skin in their faces and chests which act like bubble-wrap, cushioning the impact with water.

Predator Cooperation

The most effective strategy predators use against schooling fish is to first scare them into forming a bait ball. Strategies such as those outlined in the previous section, can work to a degree against freely streaming fish schools, but work much better if the fish school is first compacted into a bait ball. It is difficult for predators working individually to scare a fish school into a bait ball, and they usually work together in a cooperative effort.

- Thresher sharks compact their prey by swimming in circles around them, splashing the water with their long tails, often in pairs or small groups. They then strike sharply at the bait ball to stun the fish with the upper lobe of their tails.
- Schools of forage fish can draw silky sharks in large numbers. Silky sharks have been documented "herding" such schools into a bait ball trapped against the surface, and then consuming the entire school. When attacking tightly packed fish, silky sharks charge through the ball and slash open-mouthed,

catching the prey fish at the corners of their jaws. Although multiple individuals may feed at once, each launches its attack independently.

- Pods of many dolphin species commonly herd a school of fish into a bait ball while individual members take turns ploughing through and feeding on the more compacted shoal. *Corralling* is a method where fish are chased to shallow water where they are more easily captured. Some dolphins takes this further with *strand feeding*, driving prey until they are stranded on mud banks, where they can be easily accessed. Dolphins have also been observed blowing bubbles to startle and separate individual fish from a bait ball.
- The humpback whale uses a feeding technique called *bubble net feeding*. A group of whales swim in a shrinking circle blowing bubbles below a school of prey fish. Forage fish show a strong fear of bubbles and can be easily contained within a bubble curtain. The shrinking ring of bubbles encircles the school and confines it in an ever smaller cylinder. The whales then lunge feed, often as a synchronised group, suddenly swimming upward through the "net", mouths agape, swallowing thousands of fish in one gulp. The ring can begin at up to 30 metres (98 ft) in diametre with perhaps a dozen whales cooperating. Using a crittercam attached to whale's back it was discovered that some whales blow the bubbles while others dive deeper to drive fish toward the surface, and yet others herd the prey into the net by vocalising. Some humpback whales also scare schooling fish by slapping their tails (lobtail). Although many whale species lunge feed, only humpbacks use bubble nets.
- Killer whales usually hunt larger fish, such as salmon, individually or in small group of individuals. However, forage fish, such as herrings, are often caught using *carousel feeding*. The killer whales force the herring into a tight ball by releasing bursts of bubbles or flashing their white undersides. They then slap the ball with their tail flukes, either stunning or killing up to 10–15 herring with a successful slap. The herring are then eaten one at a time. Carousel feeding has only been documented in the Norwegian killer whale population and with some oceanic dolphin species.

Mixed Species Feeding

Fish that school in large numbers can draw the attentions of many different predator species. The attraction of huge numbers of prey fish means that these predator species, which might otherwise be mutually antagonistic, usually cooperate with each other in pursuit of their common goal.

Bryde's whales often track dolphin groups as they herd prey species. As soon as the dolphins corral the prey, the whales lunge feed through the centre.

In 2001, Clua and Grosvalet proposed a four stage model to describe mixed species feeding behaviour involving common dolphins, tuna and shearwater sea birds.

- In the preparation phase, the dolphins rapidly circle the fish school to compact it.
- In the intensification phase, the bait ball is further structured by dolphins circling and foraging the periphery of the ball, while shearwaters forage from the ocean surface.
- At the mature phase, the dolphins temporarily scatter as giant tuna lunge vertically through the centre of the ball. At this point the concentration of the bait ball is optimal.
- Finally in the dispersion phase, the tuna and sea birds disperse, and some dolphins return to forage on the remnants of the ball.

Most years, off southern Africa between May and July, billions of sardines (specifically the Southern African pilchard *Sardinops sagax*) spawn in the cool waters of the Agulhas Bank and move northward along the east coast of South Africa. This great fish migration is called the sardine run. Their sheer numbers create a feeding frenzy along the coastline. The run, containing millions of individual sardines, occurs when a current of cold water heads north from the Agulhas Bank up to Mozambique where it then leaves the coastline and goes further east into the Indian Ocean. During the sardine run, as many as 18,000 dolphins, behaving like sheepdogs, herd the sardines into bait balls, or corral them in shallow water. Once rounded up, the dolphins and other predators take turns ploughing through the bait balls, gorging on the fish as they sweep through. Seabirds also attack them from above, flocks of gannets, cormorants, terns and gulls. The sardine run is featured in the first episode of the 2001 BBC nature documentary *The Blue Planet* and in the 2008 3D IMAX documentary film *Wild Ocean*.

A similar great migration of herrings occurs each year during the summer plankton bloom along the coast of British Columbia and Alaska. The migration is featured in the final episode of the 2009 BBC wildlife documentary *Nature's Great Events*. In winter, the coastal fjords and inlets are relatively lifeless, and the resident Steller sea lions must dive deeper and further from the coast to catch the widely-dispersed herring. Humpback whales overwinter in the warm Pacific waters off Hawaii, where new mothers suckle their calves. They begin their 3,000 mile journey north in early spring, when the sea lions also give birth to their young. Spring storms disturb nutrients in the water which, together with the strengthening power of the sun, act as the catalysts for the plankton bloom. Huge shoals of herring arrive to spawn, turning the shallows milky white.

The herring sift plankton from the water. In their wake come larger predators, including Pacific white-sided dolphins and killer whales. Common murres dive under the herring shoals and pick off the fish from below, pinning them to the surface. Their defence is to form a bait ball, but gulls gathering on the surface attack them from above. The finale to the programme features unique underwater footage of humpbacks engulfing whole bait balls, and reveals their co-operative hunting behaviour called bubble-netting.

Bottom Feeder

A bottom feeder is an aquatic animal that feeds on or near the bottom of a body of water. The body of water could be the ocean, a lake, a river, or an aquarium. Bottom feeder is a general term which is used particularly in the context of aquariums. Biologists often use the term benthos, particularly for bottom feeding invertebrates, such as shellfish, crabs, crayfish, sea anemones, starfish, snails, bristleworms and sea cucumbers. However the term benthos includes all aquatic life that lives on or near the bottom, which means it also includes non-animals, such as plants and algae.

Biologists also use specific terms that refer to bottom feeding fish, such as demersal fish, groundfish, benthic fish and benthopelagic fish. Examples of bottom feeding fish specie groups are flatfish (halibut, flounder, plaice, sole), eels, cod, haddock, bass, grouper, carp, bream (snapper) and some species of catfish and shark.

Biological Trends

Bottom feeders are not necessarily detritivores, although there are many that are. Some bottom feeders graze along the bottom, feeding on plant material. Other bottom feeders may feed on other bottom feeders, and thus are carnivores. Many bottom feeders are also capable of burying themselves. A variety of invertebrates are able to bury themselves, such as bristleworms, sea cucumbers, and snails. Many vertebrate bottom feeders may bury themselves, such as flatfish or stingray.

In many bottom feeders, a mechanism to deal with substrate is often necessary. In the case of some organisms such as sea cucumbers, the sand is usually passed through the body. In fish, sand will be pumped out of the mouth through the gill slit.

In fish, most bottom feeders exhibit a flat ventral region so as to more easily rest their body on the substrate. The exception may be the flatfish, which are laterally depressed but lie on their sides. Also, many exhibit what is termed an "inferior" mouth, which means that the mouth is pointed downwards; this is beneficial as their food is often going to be below them in the substrate. Those bottom feeders with upward-pointing mouths, such as stargazers, tend to seize swimming prey. Some flatfish such as halibut actually have a "migrating" eye that moves to the upward-facing side of the fish as it ages.

Aquarium Care

In the aquarium, bottom feeders are popular as it is perceived that they will "clean up" the tank. Generally, they are only useful for consuming the extra (fresh) food left by overfed or clumsy livestock; the added biomass of additional organisms means that the aquarium will likely be more dirty. Some specialised bottom feeders are more specifically sold as "algae eaters" to increase the amount of free oxygen and aesthetic appeal of a tank.

Cleaner Fish

Cleaner fish are fish that provide a service to other fish species by removing dead skin and ectoparasites. This is an example of mutualism, an ecological interaction that benefits both parties involved. A wide variety of fishes have been observed to display cleaning behaviours including wrasses, cichlids, catfish, and gobies, as well as by a number of different species of cleaner shrimp. There is also at least one predatory mimic, the sabre-toothed blenny, that mimics cleaner fish but in fact feeds on healthy scales and mucous.

Diversity of Cleaner Fish

Marine Fishes: The best known cleaner fish are the cleaner wrasses of the genus *Labroides* found on coral reefs in the Indian Ocean and Pacific Ocean. These small fish maintain so-called cleaning stations where other fish, known as hosts, will congregate and perform specific movements to attract the attention of the cleaner fish. Remarkably, these small cleaner fish will safely clean large predatory fish that would otherwise eat small fish such as these. Cleaner wrasses appear to get almost all their nutrition through this cleaning service, and when maintained in aquaria rarely survive for long because they cannot obtain enough to eat.

Cleaning behaviours have been observed in a number of other fish groups. Neon gobies of the genera *Gobiosoma* and *Elacatinus* provide a cleaning service similar to the cleaner wrasses, though this time on reefs in the Western Atlantic, providing a good example of convergent evolution. Unlike the cleaner wrasses, they also eat a variety of small animals as well being cleaner fish, and generally do well in aquaria. However, the Caribbean cleaning goby (*Elacatinus evelynae*) will gladly eat scales and mucus from the host when the ectoparasites it normally feeds on are scarce, making the relationship somewhat less than mutually beneficial. The symbiosis does not break down because the abundance of these parasites varies significantly seasonally and spatially, and the overall benefit to the larger fish outweighs any cheating on the part of the smaller.

Brackish Water Fishes

An interesting example of a cleaning symbiosis has been observed between two brackish water cichlids of the genus *Etroplus* from South Asia. The small species *Etroplus maculatus* is the cleaner fish, and the much larger *Etroplus suratensis* is the host that receives the cleaning service.

Freshwater Fishes

Cleaning is notably less common in freshwater habitats than in marine habitats. One of the few examples of cleaning is juvenile Striped Raphael catfish cleaning the piscivorous *Hoplias* cf. *malabaricus*.

Mimicry

The sabre-toothed blenny *Aspidontus taeniatus* is a blenny that mimics the ritualised dance the cleaner wrasse makes when passing fish swim by. Instead of providing a useful cleaning service, however, it bites off pieces of healthy skin and scales from the host before darting away to safety.

Diel Vertical Migration

Diel vertical migration, also known as diurnal vertical migration, is a pattern of movement that some organisms living in the ocean and in lakes undertake each day. Usually organisms move up to the epipelagic zone at night and return to the mesopelagic zone of the oceans or to the hypolimnion zone of lakes during the day. The word diel comes from the Latin dies day, and means a 24-hour period.

Discovery

During World War II the Navy was taking sonar readings of the ocean when they discovered the deep scattering layer (DSL). The DSL was caused by large groupings of organisms that scattered the sonar to create a false or second bottom. The false bottom was shallower during the day and deeper during the night; this was the first recording of diel vertical migration. Once scientists started to do more research on what was causing the DSL, it was discovered that a large range of organisms were vertically migrating. Most types of plankton and some types of nekton (fish) have exhibited some type of vertical migration, although it is not always diel.

Types and Stimuli of Vertical Migration

There are two different factors that are known to play a role in vertical migration, endogenous and exogenous. Endogenous factors originate from the organism itself; sex, age, biological rhythms, etc. Exogenous factors are environmental factors acting on the organism such as light, gravity, oxygen, temperature, predator-prey interactions, etc.

Endogenous Factors

- Endogenous rhythm- An experiment was done at the Scripps Oceanographic Research Institute which kept organisms in column tanks with light/dark cycles. A few days later the light was changed to a constant low light and the organisms still displayed diel vertical migration. Thus suggestions that some type of internal response was causing the migration.

Exogenous Factors

- Light - An organisms wants to find an optimum light intensity (isolume). Whether it is no light or a large amount of light, an organism will travel to where it is most comfortable. Studies have shown that during a full moon organisms will not migrate up as far or during an eclipse they will start to migrate.
- Temperature - sometimes thermoclines can act as a barrier that an organism will not cross.
- Salinity - in areas such as the Arctic melting ice was cause a layer of freshwater which organisms cannot cross.
- Predator kairomones - a predator might release a chemical cue which could cause its prey to vertically migrate away.

Types of Vertical Migration

- Diel - This has already been mentioned and is the most common form. Organisms migrate daily, usually up to shallow waters at night and deep waters during the day. But there are other forms of vertical migration worth mentioning.

- Seasonal - Organisms are found at different depths depending on what season it is.
- Ontogentic - Organisms spend different stages of their life cycle at different depths.

Reasons to Vertically Migrate

There are a lot of hypotheses on why organisms would vertically migrate. Each of these hypotheses are equally valid and most of the time more than one of the hypothesis account for a reason a species would vertical migrate.

- Predator avoidance - Organisms come up to shallow waters at night to feed while it's dark out because their predators cannot see them as easily.
- Metabolic Advantages - By feeding in the warm surface waters at night and residing in the cooler deep waters during the day they can conserve energy. Alternatively, organisms feeding on the bottom in cold water during the day may migrate to surface waters at night in order to digest their meal at warmer temperatures.
- Dispersal and Transport - Organisms can use deep and shallow currents to find food patches or to maintain a geographical location.
- Avoid UV Damage - The sunlight can penetrate into the water column. If an organism, especially something small like a microbe, is too close to the surface the UV can actually damage them. So they would want to avoid getting too close to the surface, especially during daylight.

Importance with the Biological Pump

The biological pump is the conversion of CO2 and inorganic nutrients by plant photosynthesis into particulate organic matter in the euphotic zone and transferred to the deeper ocean. This is a major process in the ocean and without vertical migration it wouldn't be nearly as efficient.

The deep ocean gets most of its nutrients from higher up in the water column and it sinks down in the form of marine snow. Marine snow is made up of dead or dying animals and microbes, fecal matter, sand, and other inorganic material. Fecal matter is what vertical migration helps get to depth much faster.

Organisms migrate up to feed at night so when they migrate back to depth during the day they defecate large sinking fecal pellets. Whilst some larger fecal pellets can sink quite fast, the speed that organisms move back to depth is still faster. At night organisms are in the top 100 metres of the water column but during the day they move down to between 800–1000 metres. If organisms were to defecate at the surface it would take the fecal pellets days to reach the depth that they reach in a matter of hours.

Therefore by releasing fecal pellets at depth they have almost 1000 metres less to travel to get to the deep ocean. This is something known as active transport. The organisms are playing a more active role in moving organic matter down to depths. Because a large majority of the deep sea, especially marine microbes, depends on nutrients falling down, the quicker they can reach the ocean floor the better.

Zooplankton and salps are to play a large role in the active transport of fecal pellets. 15-50% of zooplankton biomass above 15-50% is estimated to vertically migrate and this can account for the transport of 5-45% of Particulate Organic Nitrogen to depth. As for salps, they are large gelatinous plankton that can vertically migrate 800 metres and eat large amounts of food at the surface. They also have a very long gut retention time so fecal pellets usually are released at maximum depth. Salps are also known for having some of the largest fecal pellets. Because of this they have a very fast sinking rate and small detritus particles are known to aggregate on them. This makes them sink that much faster. So while currently there is still a lot of research being done on why organisms vertically migrate, it is very clear that vertical migration plays a large role in the active transport of dissolved organic matter to depth.

Electric Fish

An electric fish is a fish that can generate electric fields. It is said to be *electrogenic*; a fish that has the ability to detect electric fields is said to be *electroreceptive*. Most electrogenic fish are also electroreceptive. Electric fish species can be found both in the sea and in freshwater rivers of South America (Gymnotiformes) and Africa (Mormyridae).

Many fish such as sharks, rays and catfishes can detect electric fields, and are thus electroreceptive, but as they cannot generate an electric field they are not classified as electric fish. Most common bony fish (teleosts), including most fish kept in aquaria or caught for food, are neither electrogenic nor electroreceptive.

Strongly and Weakly Electric Fish

Electric fish produce their electrical fields from a specialised structure called an electric organ. This is made up of modified muscle or nerve cells, which became specialised for producing bioelectric fields stronger than those that normal nerves or muscles produce (Albert and Crampton, 2006). Typically this organ is located in the tail of the electric fish. The electrical output of the organ is called the *electric organ discharge* (EOD).

Fish that have an EOD that is powerful enough to stun their prey are called strongly electric fish. The amplitude of the signal can range from 10 to 500 Volts with a current of up to 1 Ampere. Typical examples are the electric eel (*Electrophorus electricus*; not a true eel but a knifefish), the electric catfishes (family Malapteruridae), and electric rays (order Torpediniformes). By contrast, weakly electric fish generate a discharge that is typically less than one volt in amplitude. These are too weak to stun prey, but are used for navigation, object detection (electrolocation) and communication with other electric fish (electrocommunication). Some of the best known and most studied examples are Peters' elephantnose fish (*Gnathonemus petersi*) and the black ghost knifefish (*Apteronotus albifrons*).

The EOD waveform takes two general forms depending on the species. In some species the waveform is continuous and almost sinusoidal (for example the genera *Apteronotus*, *Eigenmannia* and *Gymnarchus*) and these are said to have a wave-type EOD. In other species, the EOD waveform consists of brief pulses separated by longer gaps (for example *Gnathonemus*, *Gymnotus*, *Raja*) and these are said to have a pulse-type EOD.

Table of Electric Fish

Following is a table of all known electric fish species within fresh water. There are two groups of marine fishes, the electric rays (Torpediniformes: Narcinidae and Torpedinidae) and the stargazers (Perciformes: Uranoscopidae) capable of generating strong electric pulses.

Filter Feeder

Filter feeders (a subgroup of suspension feeders) are animals that feed by straining suspended matter and food particles from water, typically by passing the water over a specialised filtering structure. Some animals that use this method of feeding are clams, krill, sponges,, baleen whales, and many fish and some sharks. Some birds, such as flamingos, are also filter feeders. Filter feeders can play an important role in clarifying water.

Examples

Fish: Most forage fish are filter feeders. For example, the *Atlantic menhaden*, a type of herring, lives on plankton caught in midwater. Adult menhaden can filter up to four gallons of water a minute; and play an important role in clarifying ocean water. They are also a natural check to the deadly red tide. In addition to these bony fish, four shark subclass species are also filter feeders.

- The whale shark sucks in a mouthful of water, closes its mouth and expels the water through its gills. During the slight delay between closing the mouth and opening the gill flaps, plankton is trapped against the dermal denticles which line its gill plates and pharynx. This fine sieve-like apparatus, which is a unique modification of the gill rakers, prevents the passage of anything but fluid out through the gills (anything above 2 to 3 mm in diametre is trapped). Any material caught in the filter between the gill bars is swallowed. Whale sharks have been observed "coughing" and it is presumed that this is a method of clearing a build up of food particles in the gill rakers.
- The megamouth shark has luminous organs called photophores around its mouth. It is believed they may exist to lure plankton or small fish into its mouth.
- The basking shark is a passive filter feeder, filtering zooplankton, small fish and invertebrates from up to 2,000 tons of water per hour. Unlike the megamouth and whale sharks, the basking shark does not appear to actively seek its quarry, but it does possess large olfactory bulbs that may guide it in the right direction. Unlike the other large filter feeders, it relies only on the water that is pushed through the gills by swimming; the megamouth shark and whale shark can suck or pump water through their gills.
- Manta rays, also belonging to the shark subclass, can time their arrival at the spawning of large shoals of fish and feed on the free-floating eggs and sperm. This stratagem is also employed by whale sharks.

Crustaceans

- Mysidacea are three cm long shrimps that live close to shore and hover above the sea floor, constantly collecting particles with their filter basket. They are an important food source for herring, cod, flounder, and striped bass. Mysids have a high resistance to toxins in polluted areas, and may contribute to high toxin levels in their predators.

- The Antarctic krill manages to directly utilise the minute phytoplankton cells, which no other higher animal of krill size can do. This is accomplished through filter feeding, using the krill's developed front legs, providing for a very efficient filtering apparatus: the six thoracopods form a very effective "feeding basket" used to collect phytoplankton from the open water. In the animation at the top of this page, the krill is hovering at a 55° angle on the spot. In lower food concentrations, the feeding basket is pushed through the water for over half a metre in an opened position, and then the algae are combed to the mouth opening with special setae on the inner side of the thoracopods.
- Porcelain crab species have feeding appendages covered with setae to filter food particles from the flowing water.
- All 1,220 known species of Barnacles are filter feeders, using their highly modified legs to sift plankton from the water.

Baleen Whales

The baleen whales, also called whalebone whales or great whales, form the Mysticeti, one of two suborders of the Cetacea (whales, dolphins, and porpoises). Baleen whales are characterised by having baleen plates for filtering food from water, rather than having teeth. This distinguishes them from the other suborder of cetaceans, the toothed whales or Odontoceti. The suborder contains four families and fourteen species. The scientific name derives from the Greek word mystidos, which means "unknowable".

Bivalves

Bivalves are aquatic molluscs which have two-part shells. Typically both shells (or valves) are symmetrical along the hinge line. The class has 30,000 species, including scallops, clams, oysters and mussels. Most bivalves are filter feeders (although some have taken up scavenging and predation), extracting organic matter from the sea in which they live. Nephridia, the shell fish version of kidneys, remove the waste material. Buried bivalves feed by extending a siphon to the surface.

As an example, oysters draw water in over their gills through the beating of cilia. Suspended food (phytoplankton, zooplankton, algae and other water-borne nutrients and particles) are trapped in the mucus of a gill, and from there are transported to the mouth, where they are eaten, digested and expelled as feces or pseudofeces. Each oyster filters up to five litres of water per hour. Scientists believe that the Chesapeake Bay's once-flourishing oyster population historically filtered the estuary's entire water volume of excess nutrients every three or four days. Today that process would take almost a year, and sediment, nutrients, and algae can cause problems in local waters. Oysters filter these pollutants, and either eat them or shape them into small packets that are deposited on the bottom where they are harmless.

Sponges

Sponges have no true circulatory system; instead, they create a water current which is used for circulation. Dissolved gases are brought to cells and enter the cells via simple diffusion. Metabolic wastes are also transferred to the water through diffusion. Sponges pump remarkable amounts of water. *Leuconia*, for example, is a small leuconoid sponge about 10 cm tall and 1 cm in diametre. It is estimated that water enters through more than 80,000 in current canals at a speed of 6 cm per minute.

However, because *Leuconia* has more than 2 million flagellated chambers whose combined diametre is much greater than that of the canals, water flow through chambers slows to 3.6 cm per hour. Such a flow rate allows easy food capture by the collar cells. All water is expelled through a single osculum at a velocity of about 8.5 cm/second: a jet force capable of carrying waste products some distance away from the sponge.

Flamingos

Flamingos filter-feed on brine shrimp. Their oddly-shaped beaks are specially adapted to separate mud and silt from the food they eat, and are uniquely used upside-down. The filtering of food items is assisted by hairy structures called lamellae which line the mandibles, and the large rough-surfaced tongue.

Other Filter Feeders

Other examples of filter-feeding organisms include:

- Brachiopoda (a phylum within the Eumetazoa subkingdom of animals, commonly referred to as lamp shells)
- Branchiopoda (a class within the Arthropod phylum of animals. Branchiopods are primitive crustaceans found mostly in freshwater)
- Bryozoa (a phylum within the Eumetazoa subkingdom of animals, commonly referred to as moss animals or sea mats)

- Canalipalpata (an order of Polychaete ringed worms, commonly referred to as bristle-footed annelids)
- Ctenophora (a phylum within the Eumetazoa subkingdom of animals, commonly referred to as comb jellies)
- Echiura (an order of Polychaete ringed worms, commonly referred to as spoon worms)
- Entoprocta (a phylum within the Eumetazoa subkingdom of animals, commonly referred to as goblet worms)
- Lobodon carcinophagus (a subphylum of the Chordata phylum of animals, commonly referred to as Crabeater seals)
- Holothuroidea (a class within the Echinodermata phylum of animals, commonly referred to as sea cucumbers)
- Phoronida (a phylum within the Eumetazoa subkingdom of animals, commonly referred to as horseshoe worms)
- Sipuncula (a phylum within the Eumetazoa subkingdom of animals, commonly referred to as peanut worms)
- Urochordata (a subphylum of the Chordata phylum of animals, commonly referred to as sea squirts).

Forage Fish

Forage fish, also called prey fish or bait fish, are small fish which are preyed on by larger predators for food. Predators include other larger fish, seabirds and marine mammals. Typical ocean forage fish feed near the base of the food chain on plankton, often by filter feeding. They include the fishes of the family Clupeidae (herrings, shad, sardines, hilsa, menhaden and sprats), as well as anchovies, capelin and halfbeaks.

Forage fish compensate for their small size by forming schools. Some swim in synchronised grids with their mouths open so they can efficiently filter plankton. These schools can become immense shoals which move along coastlines and migrate across open oceans. The shoals are concentrated fuel resources for the great marine predators. The predators are keenly focused on the shoals, acutely aware of their numbers and whereabouts, and make migrations themselves that can span thousands of miles to connect, or stay connected, with them.

The ocean primary producers, mainly contained in plankton, produce food energy from the sun and are the raw fuel for the ocean food webs. Forage fish transfer this energy by eating the plankton and becoming food themselves for the top predators. In this way, forage fish occupy the central positions in ocean and lake food webs.

In recent times, many of the worlds great predator fisheries have collapsed. To compensate, the fishing industry is removing huge amounts of forage fish from the oceans, using factory ships with sophisticated sonar and spotting planes. Most of the catch is fed to farmed animals. Fisheries scientists are expressing concern that this will result in further collapses of the predator fish that depend on them.

In the Oceans

Typical ocean forage fish are small, silvery schooling oily fish such as herring, anchovies and menhaden, and other small, schooling baitfish like capelin, smelts, sand lance, halfbeaks, pollock, butterfish and juvenile rockfish. Herrings are a preeminent forage fish, often marketed as sardines or pilchards.

The term "forage fish" is a term used in fisheries, and is applied also to forage species that are not true fish, but play a significant role as prey for predators. Thus invertebrates such as squid and shrimp are also referred to as "forage fish". Even the tiny shrimp-like creatures called krill, small enough to be eaten by other forage fish, yet large enough to eat the same zooplankton as forage fish, are often classified as "forage fish". Forage fish utilise the biomass of copepods, mysids and krill in the pelagic zone to become the dominant converters of the enormous ocean production of zooplankton. They are, in turn, central prey items for higher trophic levels. Forage fish may have achieved their dominance because of the way they live in huge, and often extremely fast cruising schools.

Though forage fish are abundant, there are relatively few species. There are more species of primary producers and apex predators in the ocean than there are forage fish.

Ocean food Webs

Forage fish occupy central positions in the ocean food webs. The position that a fish occupies in a food web is called its trophic level (Greek *trophç* = food). The organisms it eats are at a lower trophic level, and the organisms that eat it are at a higher trophic level. Forage fish occupy middle levels in the food web, serving as a dominant prey to higher level fish, seabirds and mammals.

Ecological pyramids are graphical representations, along the lines of the diagram at the right, which show how biomass or productivity changes at each trophic level in a ecosystem. The first or bottom level is occupied by primary producers or autotrophs (Greek *autos* = self and *trophe* = food). These are the names given to organisms that do not feed on other organisms, but produce biomass from inorganic compounds, mostly by a process of photosynthesis. In oceans, most primary production is performed by algae. This is a contrast to land, where most primary production is performed by vascular plants. Algae ranges from single floating cells to attached seaweeds, while vascular plants are represented in the ocean by groups such as the seagrasses. Larger producers, such as seagrasses and seaweeds, are mostly confined to the littoral zone and shallow waters, where they attach to the underlying substrate and still be within the photic zone. Most primary production in the ocean is performed by microscopic organisms, the phytoplankton.

Thus, in ocean environments, the first bottom trophic level is occupied principally by phytoplankton, microscopic drifting organisms, mostly one-celled algae, that float in the sea. Most phytoplankton are too small to be seen individually with the unaided eye. They can appear as a green discolouration of the water when they are present in high enough numbers. Since they increase their biomass mostly through photosynthesis they live in the sun-lit surface layer (euphotic zone) of the sea.

The most important groups of phytoplankton include the diatoms and dinoflagellates. Diatoms are especially important in oceans, where they are estimated to contribute up to 45% of the total ocean's primary production. Diatoms are usually microscopic, although some species can reach up to 2 millimetres in length.

The second trophic level (primary consumers) is occupied by zooplankton which feed off the phytoplankton. Together with the phytoplankton, they form the base of the food pyramid that supports most of the world's great fishing grounds. Zooplankton are tiny animals found with the phytoplankton in oceanic surface waters, and include tiny crustaceans, and fish larvae and fry (recently-hatched fish). Most zooplankton are filter feeders, and they use appendages to strain the phytoplankton in the water. Some larger zooplankton also feed on smaller zooplankton. Some zooplankton can jump about a bit to avoid predators, but they can't really swim. Like phytoplankton, they float with the currents, tides and winds instead. Zooplanktons can reproduce rapidly, their populations can increase up to thirty percent a day under favourable conditions. Many live short and productive lives and reach maturity quickly.

Particularly important groups of zooplankton are the copepods and krill. These are not shown in the images above, but are discussed in more detail later. Copepods are a group of small crustaceans found in ocean and freshwater habitats. They are the biggest source of protein in the sea, and are important prey for forage fish. Krill constitute the next biggest source of protein. Krill are particularly large predator zooplankton which feed on smaller zooplankton. This means they really belong to the third trophic level, secondary consumers, along with the forage fish.

Together, phytoplankton and zooplankton make up most of the plankton in the sea. Plankton is the term applied to any small drifting organisms that float in the sea (Greek *planktos* = wanderer or drifter). By definition, organisms classified as plankton are unable to swim against ocean currents; they cannot resist the ambient current and control their position. In ocean environments, the first two tropic levels are occupied mainly by plankton. Plankton are divided into producers and consumers.

The producers are the phytoplankton (Greek *phyton* = plant) and the consumers, who eat the phytoplankton, are the zooplankton (Greek *zoon* = animal).

Diet

Forage fish feed on plankton. When they are eaten by larger predators, they transfer this energy from the bottom of the food chain to the top and in this way are the central link between trophic levels.

Forage fish are usually filter feeders, meaning that they feed by straining suspended matter and food particles from water. They usually travel in large, slow moving, tightly packed schools with their mouths open. They are typically omnivorous. Their diet is usually based primarily on zooplankton, although, since they are omnivorous, they also take in some phytoplankton.

Young forage fish, such as herring, mostly feed on phytoplankton and as they mature they start to consume larger organisms. Older herrings feed on zooplankton, tiny animals that are found in oceanic surface waters, and fish larvae and fry (recently-hatched fish). Copepods and other tiny crustaceans are common zooplankton eaten by forage fish. During daylight, many forage fish stay in the safety of deep water, feeding at the surface only at night when there is less chance of predation. They swim with their mouths open, filtering plankton from the water as it passes through their gills. Ocean halfbeaks are omnivores which feed on algae, plankton, marine plants like seagrass, invertebrates like pteropods and crustaceans and smaller fishes. Some tropical species feed on animals during the day and plants at night, while others alternate summer carnivory with winter herbivory. They are in turn eaten by billfish, mackerel, and sharks.

Predators

Forage fish are the food that sustains larger predators above them in the ocean food chain. The superabundance they present in their schools make them ideal food sources for top predator fish such as tuna, striped bass, cod, salmon, barracuda and swordfish, as well as sharks, whales, dolphins, porpoises, seals, sea lions, and seabirds.

Schooling

Forage fish compensate for their small size by forming schools. These sometimes immense gatherings fuel the ocean food web. Most forage fish are pelagic fish, which means they form their schools in open water, and not on the bottom (benthic fish) or near the bottom (benthopelagic fish). They are short-lived, and go mostly unnoticed by humans, apart from an occasional support role in a documentary about a great ocean predator. While we may not pay them much attention, the great marine predators are keenly focused on them, acutely aware of their numbers and whereabouts, and make migrations that can span thousands of miles to connect with them. After all, forage fish are their food.

Herring are among the most spectacular schooling fish. They aggregate together in huge numbers. Schools have been measured at over four cubic kilometres in size, containing about four billion fish. These schools move along coastlines and traverse the open oceans. Herring schools in general have very precise arrangements which allow the school to maintain relatively constant cruising speeds. Herrings have excellent hearing, and their schools react very fast to a predator.

The herrings keep a certain distance from a moving scuba diver or cruising predator like a killer whale, forming a vacuole which can look like a doughnut from a spotter plane. The intricacies of schooling is far from fully understood, especially the swimming and feeding energetics. Many hypotheses to explain the function of schooling have been suggested, such as better orientation, synchronised hunting, predator confusion and reduced risk of being found. Schooling also has disadvantages, such as excretion buildup in the breathing media and oxygen and food depletion. The way the fish array in the school probably gives energy saving advantages, though this is controversial.

On calm days, schools of herring can be detected at the surface a mile away by little waves they form, or from several metres at night when they trigger bioluminescence in surrounding plankton. Underwater recordings show herring constantly cruising at high speeds up to 108 cm per second, with much higher escape speeds.

They are fragile fish, and because of their adaptation to schooling behaviour they are rarely displayed in aquaria. Even with the best facilities aquaria can offer they become sluggish compared to their quivering energy in wild schools.

Hunting Copepods

Copepods are a group of small crustaceans found in ocean and freshwater habitats. Many species are planktonic (drifting in the ocean water), while others are benthic (living on the sea floor). Copepods are typically one millimetre (0.04 in) to two millimetres (0.08 in) long, with a teardrop shaped body. Like other crustaceans they have an armoured exoskeleton, but they are so small that this armour, and the entire body, is usually transparent.

Copepods are usually the dominant zooplankton. Some scientists say they form the largest animal biomass on the planet. The other contender is the Antarctic krill. But copepods are smaller than krill, with faster growth rates, and they are more evenly distributed throughout the oceans. This means copepods almost certainly contribute more secondary production to the world's oceans than krill, and perhaps more than all other groups of marine organisms together. They are a major item on the forage fish menu.

Copepods are very alert and evasive. They have large antennae. When they spread their antennae they can sense the pressure wave from an approaching fish and jump with great speed over a few centimetres.

Herrings are pelagic feeders. Their prey consists of a wide spectrum of phytoplankton and zooplankton, amongst which copepods are the dominant prey. Young herring usually capture small copepods by hunting them individually— they approach them from below. The (half speed) video loop at the left shows a juvenile herring feeding on copepods. In the middle of the image a copepod escapes successfully to the left. The opercula (hard bony flaps covering the gills) are spread wide open to compensate the pressure wave which would alert the copepod to trigger a jump.

If prey concentrations reach very high levels, the herrings adopt a method called "ram feeding". They swim with their mouth wide open and their opercula fully expanded. Every several feet, they close and clean their gill rakers for a few milliseconds (filter feeding). In the photo on the right, herring ram feed on a school of copepods. The fish all open their mouths and opercula wide at the same time. The fish swim in a grid where the distance between them is the same as the jump length of their prey, as indicated in the animation below. In the animation, juvenile herring hunt the copepods in synchronisation: The copepods sense with their antennae the pressure-wave of an approaching herring and react with a fast escape jump. The length of the jump is fairly constant. The fish align themselves in a grid with this characteristic jump length. A copepod can dart about 80 times before it tires out. After a jump, it takes it 60 milliseconds to spread its antennae again, and this time delay becomes its undoing, as the almost endless stream of herrings allows a herring to eventually snap the copepod. A single juvenile herring could never catch a large copepod.

Migrations

Forage fish often make great migrations between their spawning, feeding and nursery grounds. Schools of a particular stock usually travel in a triangle between these grounds. For example, one stock of herrings have their spawning ground in southern Norway, their feeding ground in Iceland, and their nursery ground in northern Norway. Wide triangular journeys such as these may be important because forage fish, when feeding, cannot distinguish their own offspring.

Fertile feeding grounds for forage fish are provided by ocean upwellings. Oceanic gyres are large-scale ocean currents caused by the Coriolis effect. Wind-driven surface currents interact with these gyres and the underwater topography, such as seamounts and the edge of continental shelves, to produce downwellings and upwellings. These can transport nutrients which plankton thrive on. The result can be rich feeding grounds attractive to the plankton feeding forage fish. In turn, the forage fish themselves become a feeding ground for larger predator fish. Most upwellings are coastal, and many of them support some of the most productive fisheries in the world. Regions of notable upwelling include coastal Peru, Chile, Arabian Sea, western South Africa, eastern New Zealand and the California coast.

Capelin are a forage fish of the smelt family found in the Atlantic and Arctic oceans. In summer, they graze on dense swarms of plankton at the edge of the ice shelf. Larger capelin also eat krill and other crustaceans. The capelin move inshore in large schools to spawn and migrate in spring and summer to feed in plankton rich areas between Iceland, Greenland, and Jan Mayen. The migration is affected by ocean currents. Around Iceland maturing capelin make large northward feeding migrations in spring and summer. The return migration takes place in September to November. The spawning migration starts north of Iceland in December or January.

The diagram on the right shows the main spawning grounds and larval drift routes. Capelin on the way to feeding grounds is coloured green, capelin on the way back is blue, and the breeding grounds are red. In a paper published in 2009, researchers from Iceland recount their application of an interacting particle model to the capelin stock around Iceland, successfully predicting the spawning migration route for 2008.

Predator Attacks

Schooling forage fish are subject to constant attacks by predators. An example is the attacks that take place during the African sardine run. The African sardine run is a spectacular migration by millions of silvery sardines along the southern coastline of Africa. In terms of biomass, the sardine run could rival East Africa's great wildebeest migration.

Sardines have a short life-cycle, living only two or three years. Adult sardines, about two years old, mass on the Agulhas Bank where they spawn during spring and summer, releasing tens of thousands of eggs into the water.

The adult sardines then make their way in hundreds of shoals towards the subtropical waters of the Indian Ocean. A larger shoal might be 7 kilometres (4 mi) long, 1.5 kilometres (1 mi) wide and 30 metres (100 ft) deep. Huge numbers of sharks, dolphins, tuna, sailfish, Cape fur seals and even killer whales congregate and follow the shoals, creating a feeding frenzy along the coastline.

When threatened, sardines instinctively group together and create massive "bait balls". Bait balls can be up to 20 metres (70 ft) in diametre. They are short lived, seldom lasting longer than 20 minutes. As many as 18,000 dolphins, behaving like sheepdogs, round the sardines into these bait balls, or herd them to shallow water (corralling) where they are easier to catch. Once rounded up, the dolphins and other predators take turns plowing through the bait balls, gorging on the fish as they sweep through. Seabirds also attack them from above, flocks of gannets, cormorants, terns and gulls. Some of these seabirds plummet from heights of 30 metres (100 feet), plunging through the water leaving vapour-like trails behind like fighter planes, .

The eggs, left behind at the Agulhas Banks, drift northwest with the current into waters off the west coast, where the larvae develop into juvenile fish. When they are old enough, they aggregate into dense shoals and migrate southwards, returning to the Agulhas banks in order to restart the cycle.

Forage Fisheries

Herring has been known as a staple food source since 3000 B.C. In Roman times, anchovies were the base for the fermented fish sauce called *garum*. This staple of cuisine was produced in industrial quantities and transported over long distances.

Fishing for sardela or sardina (*Sardina pilchardus*) is an ongoing activity on the Croatian Adriatic coasts of Dalmatia and Istria. It traces its roots back thousands of years. The region was then largely a Venetian dominion, part of the Roman Empire. The area has always been sustained through fishing mainly sardines. Along the coast towns still promote the traditional practice of fishing by lateen sail boats for tourism and festivals.

Pilchard fishing and processing thrived in Cornwall between 1750 and 1880, after which stocks went into an almost terminal decline. Recently (2007) stocks have been improving. The industry has featured in many works of art, including Stanhope Forbes and other Newlyn School artists.

Contemporary

Traditional commercial fisheries were directed towards high value ocean predators such as cod, rockfish and tuna, rather than humble forage fish. As technologies developed, fisheries became so effective at locating and catching predator fish that many of the stocks collapsed. The industry compensated by turning to species lower in the food chain.

In former times, forage fish were more difficult to fish profitably, and were a small part of the global marine fisheries. But modern industrial fishing technologies have enabled the removal of increasing quantities. Industrial-scale forage fish fisheries need large scale landings of fish to return profits. They are dominated by a small number of corporate fishing and processing companies.

Forage fish populations are very vulnerable when faced with modern fishing equipment. They swim near the surface in compacted schools, so they are relatively easy to locate at the surface with sophisticated electronic fishfinders and from above with spotter planes. Once located, they are scooped out of the water using highly efficient nets, such as purse seines, which remove most of the school.

Spawning patterns in forage fish are highly predictable. Some fisheries use knowledge of these patterns to harvest the forage species as they come together to spawn, removing the fish before they have actually spawned.

Fishing during spawning periods or at other times when forage fish amass in large numbers can also be a blow to predators. Many predators, such as whales, tuna and sharks, have evolved to migrate long distances to specific sites for feeding and breeding. Their survival hinges on their finding these forage schools at their feeding grounds.

The great ocean predators find that, no matter how they are adapted for speed, size, endurance or stealth, they are on the losing side when faced with the machinery of contemporary industrial fishing.

Altogether, forage fish account for 37 percent (31.5 million tonnes) of all fish taken from the world's oceans each year. However, because there are fewer species of forage fish compared to predator fish, forage species fisheries are the largest in the world. Seven of the top ten fisheries target forage fish. The total world catch of herrings, sardines and anchovies alone in 2005 was 22.4 million tonnes, 24 percent of the total world catch.

The Peruvian anchoveta fishery is now the biggest in the world (10.7 million tonnes in 2004), while the Alaskan pollock fishery in the Bering Sea is the largest single species fishery in the world (3 million tonnes). The Alaskan pollock is said to be the largest remaining single species source of palatable fish in the world.

However, the biomass of pollock has declined in recent years, perhaps spelling trouble for both the Bering Sea ecosystem and the commercial fishery it supports. Acoustic surveys by NOAA indicate that the 2008 pollock population is almost 50 percent lower than last year's survey levels. Some scientists think this decline in Alaska pollock could repeat the collapse experienced by Atlantic cod, which could have negative consequences for the entire Bering Sea ecosystem. Salmon, halibut, endangered Steller sea lions, fur seals, and humpback whales eat pollock and depend on healthy populations to sustain themselves.

Use as Animal Feed

Eighty percent of the forage fish caught are fed to animals. Ninety percent is processed into fishmeal and fish oil. Of this, 46 percent was fed to farmed fish, 24 percent to pigs, and 22 percent to poultry (2002). Six times the weight of forage fish is fed to pigs and poultry alone than the entire seafood consumption of the U.S. market.

According to Turchini and De Silva (2008), another 2.5 million tonnes of the annual forage fish catch is consumed by the global cat food industry. In Australia, pet cats eat 13.7 kilograms of fish a year compared to the 11 kilograms eaten by the average Australian. The pet food industry is increasingly marketing premium and super-premium products, when different raw materials, such as the by-products of the fish filleting industry, could be used instead.

Environmental Issues

A recent study (2008) by fisheries scientists Jacqueline Alder, Daniel Pauly and colleagues is the product of a nine-year Sea Around Us Project.

The study concludes that...

1. The composition of landings of forage fish fisheries have changed over the past 50 years with the trophic level of fish used in fishmeal increasing over the past 20 years.
2. Our understanding of the role of forage fish in marine ecosystem and the impact of fishing is still limited.
3. Landing of forage fish peaked by the 1970s, and these high levels are highly unlikely in the future, even if fisheries are managed sustainably.
4. The consumption of forage fish by seabirds and marine mammals is not likely to be onerous to fisheries, except in a few localised areas.

   By contrast, fisheries, by reducing the biomass of small pelagics, might pose a threat to these predators, particularly to those species for which stocks have been heavily depleted by human exploitation in the past.
5. Some forage fish species are consumed by many people with consumption patterns changing over the last 20 years.

Aquaculture continues to increase its consumption of fishmeal and fish oil.

In Lakes and Rivers

Forage fish also inhabit freshwater habitats, such as lakes and rivers, where they serve as food for larger freshwater predators. Usually smaller than 15 centimetres (6 in) in length, these small bait fish make up most of the fish found in lakes and rivers.

The minnow family alone, consisting of minnows, chubs, shiners and daces, consists of more than fifty species. Other freshwater forage fish include suckers, killifish, shad, bony fish as well as fish of the sunfish family, excluding black basses and crappie, and smaller species of the carp family. There are also anadromous forage fish, such as eulachon.

Within any fresh or saltwater ecosystem, there will always be both desirable and undesirable fishes, and this varies from country to country, and often from region to region within a country. Sport fishermen divide freshwater predators of forage fish into those:

- which have a good fighting ability and are good to eat, called sport (or game) fish.
- the other less desirable fish, called rough fish in North America and coarse fish in Britain

Rough or coarse fish usually refers to fish that are not commonly eaten, not sought after for sporting reasons, or have become invasive species reducing the populations of desirable fish. They compete for forage fish with the more popular sport fish. They are often regarded as a nuisance, and are not usually protected by game laws. Forage fish generally are not considered rough or coarse fish because of their usefulness as bait.

The term *rough fish* is used by U.S. state agencies and anglers to describe undesirable predator fish. In North America, anglers fish for salmon, trout, bass, pike, catfish, walleye and muskellunge. The smallest fish are called panfish, because they can fit in a standard cooking pan. Some examples are crappies, rock bass, perch, bluegill and sunfish.

The term *coarse fish* originated in the United Kingdom in the early 19th century. Prior to that time, recreational fishing was the sport of the gentry, who angled for trout and salmon which they called "game fish". Fish other than game fish were disdained as "coarse fish".

These days, "game fish" refers to Salmonids (other than grayling) — that is, salmon, trout and char. Coarse fish are made up mostly of the larger species of Cyprinids (carp, roach, bream) as well as pike, catfish, gar and lamprey. Coarse fish are no longer disdained; indeed, fishing for coarse fish has become a popular pastime.

Bait and Feeder Fish

Forage fish are sometimes referred to as *bait fish* or *feeder fish.* Bait fish is a term used particularly by recreational fishermen, although commercial fisherman also catch fish to bait longlines and traps. Forage fish is a fisheries term, and is used in the context of fisheries.

Bait fish, by contrast, are fish that are caught by humans to use as bait for other fish. The terms overlap in the sense that most bait fish are also forage fish, and vice versa. Feeder fish is a term used particularly in the context of fish aquariums. It refers essentially to the same concept as forage fish, small fish that are eaten by larger fish, but the term is adapted to the particular requirements of working with fish in aquariums.

Timeline

- 2006: The U.S. National Coalition for Marine Conservation asks U.S. fishery managers to put "Forage First!". Their campaign was launched with the publication of their report, *Taking the Bait: Are America's Fisheries Out-competing Predators for their Prey?*, available at cost to the U. S. fishing industry, encouraging fishery managers to protect predator-prey relationships as a first step toward an ecosystem based approach to fishery management.
- 2009: The international Lenfest Forage Fish Task Force is established to develop workable management plans for tackling the depletion of forage fish.

Chapter 5

Behaviour of Birds

Bird Vocalisation

Bird vocalisation includes both bird calls and bird songs. In non-technical use, bird songs are the bird sounds that are melodious to the human ear. In ornithology and birding, (relatively complex) songs are distinguished by function from (relatively simple) calls.

Figure: *A male Blackbird (Turdus merula) singing. Bogense havn, Funen, Denmark.*

Definition

Bird song is best developed in the order Passeriformes. Most song is emitted by male rather than female birds. Song is usually delivered from prominent perches although some species may sing when flying. Some groups are nearly voiceless, producing only percussive and rhythmic sounds, such as the storks, which clatter their bills. In some manakins (Pipridae), the males have evolved several mechanisms for mechanical sound production, including mechanisms for stridulation not unlike those found in some insects.

The production of sounds by mechanical means as opposed to the use of the syrinx has been termed variously *instrumental music* by Charles Darwin, *mechanical sounds* and more recently *sonation*. The term *sonate* has been defined as the act of producing non-vocal sounds that are intentionally modulated communicative signals, produced using non-syringeal structures such as the bill, wings, tail, feet and body feathers.

Anatomy

The avian vocal organ is called the syrinx; it is a bony structure at the bottom of the trachea (unlike the larynx at the top of the mammalian trachea). The syrinx and sometimes a surrounding air sac resonate to sound waves that are made by membranes past which the bird forces air. The bird controls the pitch by changing the tension on the membranes and controls both pitch and volume by changing the force of exhalation. It can control the two sides of the trachea independently, which is how some species can produce two notes at once.

Function

Scientists hypothesize that bird song has evolved through sexual selection, and experiments suggest that the quality of bird song may be a good indicator of fitness. Experiments also suggest that parasites and diseases may directly affect song characteristics such as song rate, which thereby act as reliable indicators of health. The song repertoire also appears to indicate fitness in some species. The ability of male birds to hold and advertise territories using song also demonstrates their fitness.

Communication through bird calls can be between individuals of the same species or even across species. Birds communicate alarm through vocalisations and movements that are specific to the threat, and bird alarms can be understood by other animal species, including other birds, in order to identify and protect against the specific threat. Mobbing calls are used to recruit individuals in an area where an owl or other predator may be present. These calls are characterised by wide frequency spectra, sharp onset and termination, and repetitiveness which are common across species and are believed to be helpful to other potential "mobbers" by being easy to locate. The alarm calls of most species, on the other hand, are characteristically high-pitched making the caller difficult to locate.

Individual birds may be sensitive enough to identify each other through their calls. Many birds that nest in colonies can locate their chicks using their calls. Calls are sometimes distinctive enough for individual identification even by human researchers in ecological studies.

Many birds engage in duet calls. In some cases the duets are so perfectly timed as to appear almost as one call. This kind of calling is termed antiphonal duetting. Such duetting is noted in a wide range of families including quails, bushshrikes, babblers such as the scimitar babblers, some owls and parrots. In territorial songbirds, birds are more likely to countersign when they have been aroused by simulated intrusion into their territory. This implies a role in intraspecies aggressive competition.

Some birds are excellent vocal mimics. In some tropical species, mimics such as the drongos may have a role in the formation of mixed-species foraging flocks. Vocal mimicry can include conspecifics, other species or even man-made sounds. Many hypotheses have been made on the functions of vocal mimicry including suggestions that they may be involved in sexual selection by acting as an indicator of fitness, help brood parasites, protect against predation but strong support is lacking for any function. Many birds, and especially those that nest in cavities, are known to produce a snake like hissing sound that may help deter predators at close range.

Some cave-dwelling species, including Oilbird and Swiftlets (*Collocalia* and *Aerodramus* spp.), use audible sound (with the majority of sonic location occurring between 2 and 5 kHz) to echolocate in the darkness of caves. The only bird known to make use of infrasound (at about 20 Hz) is the western capercaillie.

The hearing range of birds is from below 50 Hz (infrasound) to above 20 kHz (ultrasound) with maximum sensitivity between 1 and 5 kHz. The range of frequencies at which birds call in an environment varies with the quality of habitat and the ambient sounds. It has been suggested that narrow bandwidths, low frequencies, low-frequency modulations, and long elements and inter-element intervals should be found in habitats with complex vegetation structures (which would absorb and muffle sounds) while high frequencies, broad bandwidth, high-frequency modulations (trills), and short elements and inter-elements may be expected in habitats with herbaceous cover. It has been hypothesized that the available frequency range is partitioned and birds call so that overlap between different species in frequency and time is reduced. This idea has been termed the "acoustic niche".

Birds sing louder and at a higher pitch in urban areas, where there is ambient low-frequency noise.

Bird Language

The language of the birds has long been a topic for anecdote and speculation. That calls have meanings that are interpreted by their listeners has been well demonstrated. Domestic chickens have distinctive alarm calls for aerial and ground predators, and they respond to these alarm calls appropriately. However a language has, in addition to words, structures and rules. Studies to demonstrate the existence of language have been difficult due to the range of possible interpretations. Research on parrots by Irene Pepperberg is claimed to demonstrate the innate ability for grammatical structures, including the existence of concepts such as nouns, adjectives and verbs. Studies on starling vocalisations have also suggested that they may have recursive structures.

The term "bird language" may also more informally refer to patterns in bird vocalisations that communicate information to other birds or other animals in general. Wilderness Awareness School groups bird vocalisations into five different classes, sometimes called "voices," each of which communicates different information. Song has been discussed at length in this article. Companion calling is a short vocalisation made between mates, parent and young, or members of a flock to maintain contact when out of visual range. Juvenile begging is a strident, loud vocalisation often made by young to a parent when begging for food. Intraspecific aggression can consist of loud, alarmed-sounding vocalisations or of energetic song, and may be heard when members of the same species behave aggressively toward each other. Alarm may be heard when birds are startled, frightened, or terrified for their lives, and can take many forms. Mobbing is one example of alarm, while a high-pitched alarm call is another.

Of the five voices of the birds, four of them communicate the message that the bird feels safe. Birds that engage in song, companion calling, juvenile begging, and intraspecific aggression all display what Jon Young calls "baseline" behaviour, or a relaxed state free of the fear of predation. Alarm communicates the presence of a predator, or an influence that the bird may see as predatory such as a human hiker. Alarms have distinct sounds and shapes, each of which is specific to the source of the disturbance. For example, ravens mobbing a hawk or owl in a tree will clump around the predator in a loose ball, calling and diving. If the ravens rise off the tree and fly higher, the predator was a hawk and has flown up to escape, as is typical of hawks. If the ravens drop out of the tree and fly low and away, the predator was an owl and has dropped low off its perch to escape, as is typical of owls.

Neurophysiology

The main brain areas involved in bird song are:

- Anterior forebrain pathway (vocal learning): composed of the lateral part of the magnocellular nucleus of anterior neostriatum (LMAN), which is a homologue to mammalian basal ganglia); Area X, which is part of the basal ganglia; and the Dorso-Lateral division of the Medial thalamus (DLM).
- Song production pathway: composed of the HVC (sometimes, inaccurately, called the hyperstriatum ventrale, pars caudalis); robust nucleus of the arcopallium (RA); and the tracheosyringeal part of the hypoglossal nucleus (nXIIts).

Both pathways show sexual dimorphism, with the male producing song most of the time. It has been noted that injecting testosterone in non-singing female birds can induce growth of the HVC and thus production of song.

Birdsong production is generally thought to start at the nucleus uvaeformis of the thalamus with signals emanating along a pathway that terminates at the syrinx. The pathway from the thalamus leads to the interfacial nucleus of the nidopallium to the HVC, and then to RA, the dorso-lateral division of the medial thalamus and to the tracheosyringeal nerve.

The gene FOXP2, defects of which affect both speech and comprehension of language in humans, becomes more active in the striatal region of songbirds during the time of song learning.

Recent research in birdsong learning has focused on the Ventral Tegmental Area (VTA), which sends a dopamine input to the para-olfactory lobe and Area X, LMAN and the ventrolateral medulla. Other researchers have explored the possibility that HVc is responsible for syllable production, while the robust nucleus of the arcopallium, the primary song output nucleus, may be responsible for syllable sequencing and production of notes within a syllable.

Learning

The songs of different species of birds vary, and are more or less characteristic of the species. In modern-day biology, bird song is typically analysed using acoustic spectroscopy. Species vary greatly in the complexity of their songs and in the number of distinct kinds of song they sing (up to 3000 in the Brown Thrasher); in some species, individuals vary in the same way. In a few species such as starlings and mockingbirds, songs imbed arbitrary elements learned in the individual's lifetime, a form of mimicry (though maybe better called "appropriation" [Ehrlich *et al.*], as the bird does not pass for another species).

As early as 1773 it was established that birds learnt calls and cross-fostering experiments were able to force a Linnet *Acanthis cannabina* to learn the song of a skylark *Alauda arvensis*. In many species it appears that although the basic song is the same for all members of the species, young birds learn some details of their songs from their fathers, and these variations build up over generations to form dialects.

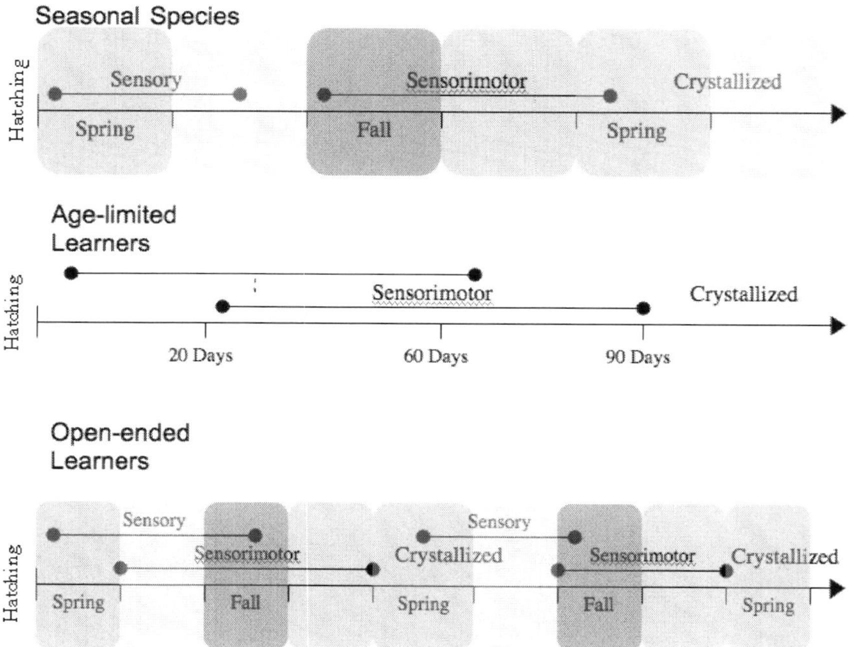

Figure: *Timeline for song learning in different species. Diagram adapted from Brainard & Doupe, 2002*

Birds learn songs early in life with sub-vocalisations that develop into renditions of adult songs. Zebra Finches, the most popular species for birdsong research, develop a version of a familiar adult's song after 20 or more days from hatch. By around 35 days, the chick will have learned the adult song. The early song is "plastic" or variable and it takes the young bird two or three months to perfect the "crystallised" song (which is less variable) of sexually mature birds.

Research indicates birds' acquisition of song is a form of motor learning that involves regions of the basal ganglia. Models of bird-song motor learning are sometimes used as models for how humans learn speech. In some species such as zebra finches, learning of song is limited to the first year; they are termed 'age-limited' or 'close-ended' learners. Other species such as the canaries can develop new songs even as sexually mature adults; these are termed 'open-ended' learners.

Researchers have hypothesized that learned songs allow the development of more complex songs through cultural interaction, thus allowing intraspecies dialects that help birds stay with their own kind within a species, and it allows birds to adapt their songs to different acoustic environments.

Auditory Feedback in Bird Song Learning

Early experiments by Thorpe in 1954 showed the importance of a bird being able to hear a tutor's song. When birds are raised in isolation, away from the influence of conspecific males, they still sing. While the song they produce resembles the song of a wild bird, it lacks the complexity and sounds distinctly different. The importance of the bird being able to hear himself sing in the sensorimotor period was later discovered by Konishi. Birds deafened before the song crystallisation period went on to produce very different songs from the wild type. These findings led scientists to believe there could be a specific part of the brain dedicated to this specific type of learning.

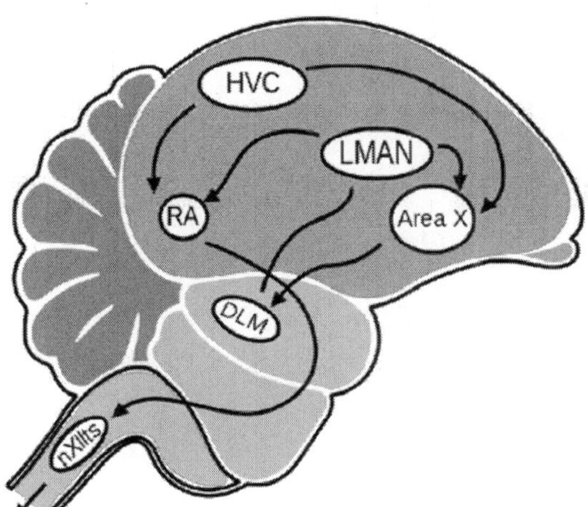

Figure: *Song learning pathway in birds (Based on Nottebohm, 2005)*

The main focus in the search for the neuronal aspect of bird song learning was guided by the song template hypothesis. This hypothesis is the idea that when a bird is young he memorizes the song of his tutor. Later, during the development phase as an adult, he matches his own trial vocalisations using auditory feedback to an acoustic template in the brain. Based on this information, he adjusts his song if needed. To find this "song template," experimenters lesioned certain parts of the brain and observed the effects.

- Lesioning the song production pathway (RA, xXII or HVc) in the brain creates serious effects on song production in all birds.
- Lesions parts of the anterior forebrain pathway, or vocal learning pathway, DLM and area X, result in deficits in learning in all birds.
- Lesioning LMAN, located in the anterior forebrain pathway in young birds disrupts song production.
- Lesioning LMAN on an adult bird shows no effect.
- Lesioning LMAN on an adult canary (an "open-ended learner" species, which can learn songs later in life) shows a progressive deterioration of song.

These results show that the area known as LMAN is the only brain area in the pathway that shows some plasticity and further studies have shown that this area of the brain responds best to the bird's own song. This neuroplasticity is vital for a bird being able to learn a song. The ability to make small adjustments based on auditory feedback is needed for the complexity of these beautiful songs. Just like any musician, birds need to practice and be able to evaluate what their song sounds like and what it's supposed to sound like in order to get it right.

To complete the picture on bird song learning, experimenters needed to discover the true plasticity of the brain. While deafening and creating auditory isolation were good techniques for discovering basic characteristics about the brain, a reversible procedure was needed to investigate further. The solution was found in disruption of the auditory feedback, or what a bird hears. A computer is able to capture the song of a singing bird and play back portions of its song, or selectively play back a certain syllable while the bird is singing. The computer is basically playing the age old trick of repeating whatever the bird sings, the "stop copying me" game.

This creates such a disruption that an adult bird will start to decrystallise its song, which includes a loss of spectral and temporal rigidity characteristic of adult song. It reverts back to the song it started singing with, before any learning took place. Furthermore, when the feedback was stopped, the birds slowly recovered their original song, something that was unheard of. These results show that there is a fair amount of plasticity retained in the brain, even for close-ended learners.

This new found plasticity in adult birds and the results on the plasticity of LMAN (shown above) combine into a model for bird song learning (diagram coming soon).

Mirror Neurons and Vocal Learning

A mirror neuron is a neuron that discharges both when an individual performs an action, and when he perceives that same action being performed by another. These neurons were first discovered in macaque monkeys, but recent research suggests that mirror neuron systems may be present in other animals including humans.

Mirror neurons have the following characteristics:

- They are located the premotor cortex
- They exhibit both sensory and motor properties
- They are action-specific – mirror neurons are only active when an individual is performing or observing a certain type of action (e.g.: grasping an object).

Because mirror neurons exhibit both sensory and motor activity, some researchers have suggested that mirror neurons may serve to map sensory experience onto motor structures. This has implications for birdsong learning– many birds rely on auditory feedback to acquire and maintain their songs. Mirror neurons may be mediating this comparison of what the bird hears and what he produces.

In search of these auditory-motor neurons, Jonathan Prather and other researchers at Duke University recorded the activity of single neurons in the HVCs of swamp sparrows. They discovered that the neurons that project from the HVC to Area X (HVC_X neurons) are highly responsive when the bird is hearing a playback of his own song. These neurons also fire in similar patterns when the bird is singing that same song. Swamp sparrows employ 3-5 different song types, and the neural activity differs depending on which song is heard or sung. The HVC_X neurons only fire in response to the presentation (or singing) of one of the songs, the primary song type. They are also temporally selective, firing at a precise phase in the song syllable.

Because the timing of the neural response is identical regardless of whether the bird was listening or singing, how can we be sure that the bird isn't just hearing himself? Prather et al. found that during the short period of time before and after the bird sings, his HVC_X neurons become insensitive to auditory input. In other words, the bird becomes "deaf" to his own song. This suggests that these neurons are producing a corollary discharge, which would allow for direct comparison of motor output and auditory input. This may be the mechanism underlying learning via auditory feedback.

Overall, the HVC_X auditory-motor neurons in swamp sparrows are very similar to the visual-motor mirror neurons discovered in primates. Like mirror neurons, the HVC_X neurons:

- Are located in a premotor brain area
- Exhibit both sensory and motor properties
- Are action-specific – a response is only triggered by the 'primary song type'

The function of the mirror neuron system is still unclear. Some scientists speculate that mirror neurons may play a role in understanding the actions of others, imitation, theory of mind and language acquisition, though there is currently insufficient neurophysiological evidence in support of these theories. Specifically regarding birds, it is possible that the mirror neuron system serves as a general mechanism underlying vocal learning, but further research is needed. In addition to the implications for song learning, the mirror neuron system could also play a role in territorial behaviours such as song-type matching and countersinging.

Identification and Systematics

The specificity of bird calls has been used extensively for species identification. The calls of birds have been described using words or nonsense syllables, or line diagrams. Common terms in English include words such as *quack*, *chirp* and *chirrup*. These are subject to imagination and vary greatly; a well-known example is the White-throated Sparrow's song, given in Canada as *O sweet Canada Canada Canada* and in New England as *Old Sam Peabody Peabody Peabody* (also *Where are you Frederick Frederick Frederick?*). In addition to nonsense words, grammatically correct phrases have been constructed as likenesses of the vocalisations of birds. For example, the Barred Owl produces a motif which some bird guides describe as *Who cooks for you? Who cooks for you all?* with the emphasis placed on *you*.

The use of spectrograms to visualize bird song was first introduced by W. H. Thorpe. These visual representations are also called sonograms or sonagrams. Some recent field guides for birds use sonograms to document the calls and songs of birds. The sonogram is objective, unlike descriptive phrases, but proper interpretation requires experience. Sonograms can also be roughly converted back into sound.

Bird song is an integral part of bird courtship and is a pre-zygotic isolation mechanism involved in the process of speciation. Many allopatric subspecies show differences in calls. These differences are sometimes minute, often detectable only in the sonograms. Song differences in addition to other taxonomic attributes have been used in the identification of new species. The use of calls has led to proposals for splitting of species complexes such as those of the *Mirafra* Bushlarks.

Recording

The first known recording of birdsong was made in 1889 by Ludwig Koch, who went on to become an eminent wildlife recordist and BBC natural history presenter.

Other notable birdsong recordists include Eric Simms and Chris Watson.

Bird Song and Music

Some musicologists believe that birdsong has had a large influence on the development of music. Although the extent of this influence is impossible to gauge, it is sometimes easy to see some of the specific ways composers have integrated birdsong with music.

There seem to be three general ways musicians or composers can be affected by birdsong: they can be influenced or inspired (consciously or unconsciously) by birdsong, they can include intentional imitations of bird song in a composition, or they can incorporate recordings of birds into their works.

In his book 'Why Birds Sing' David Rothernberg claims that birds vocalise traditional scales used in human music, such as the pentatonic scale (e.g. Hermit Thrush) and diatonic scale (e.g. Wood Thrush), providing evidence that birdsong not only sounds like music, but is music in the human sense. This claim has been refuted by Sotorrio ('Tone Spectra') who has shown that birds are not selecting 'scale tones' from a miriad of tonal possibilities, but are 'filtering out' and reinforcing the available set of overtones from the fundamental tones of their vocal chords. This requires "far less musical intelligence and deliberate appropriation" and in this regard, he suggests birdsong has something in common with Mongolian throat-singing and jaw-harp music. Sotorrio also claims that musicians like Rothernberg are deceived by "a perculiar form of Pareidolia" whereby complex tonal information is reduced to human scale concepts due to a "fixation on music as it is written rather than as it sounds". Rothernberg's claims were expored in the BBC documentary 'Why Birds Sing'.

Compositions That Imitate or use Birdsong

One early example of a composition that imitates birdsong is Janequin's "Le Chant Des Oiseaux", written in the 16th century.

Other composers who have quoted birds or have used birdsong as a compositional springboard include Vivaldi (*Spring* from the *Four Seasons*)), Biber (*Sonata Representativa*), Beethoven (*Sixth Symphony*), Wagner (*Siegfried*) and the jazz musicians Paul Winter (*Flyway*) and Jeff Silverbush (*Grandma Mickey*).

The twentieth-century French composer Olivier Messiaen composed with birdsong extensively. His Catalogue d'Oiseaux is a seven-book set of solo piano pieces based upon birdsong. His orchestral piece *Réveil des Oiseaux* is composed almost entirely of birdsong. Many of his other compositions, including *Quatuor pour la fin du temps,* similarly integrate birdsong.

The Italian composer Ottorino Respighi, with his *The Pines of Rome* (1923–1924), may have been the first to compose a piece of music that calls for pre-recorded birdsong. A few years later, Respighi wrote *Gli Uccelli* ("The birds"), based on Baroque pieces imitating birds.

The Finnish composer Einojuhani Rautavaara in 1972 wrote an orchestral piece of music called *Cantus Arcticus* (Opus 61, dubbed *Concerto for Birds and Orchestra*) making extensive use of pre-recorded birdsongs from Arctic regions, such as migrating swans.

The American jazz musician Eric Dolphy sometimes listened to birds while he practiced flute. He claimed to have incorporated bird song into some of his improvisational music.

In the 1960s and 1970s, many rock bands included sound effects in their recordings. Birds were a popular choice. The English band Pink Floyd included bird sound effects in many of the songs from their 1969 albums *Soundtrack from the Film More* and *Ummagumma* (for example, Grantchester Meadows). Similarly, the English singer Kate Bush incorporated bird sound effects into much of the music on her 2005 album, Aerial.

The Music hall artist Ronnie Ronalde has gained notoriety for his whistling imitations of birds and for integrating birdsong with human song. His songs 'In A Monastery Garden' and 'If I Were A Blackbird' include imitations of the blackbird, his "signature bird."

The French composer François-Bernard Mâche has been credited with the creation of zoomusicology, the study of the music of animals. His essay *Musique, mythe, nature, ou les Dauphins d'Arion* (1983) includes a study of "or nitho-musicology", in which he speaks of "animal musics" and a longing to connect with nature.

The German DJ, techno music producer and naturalist Dominik Eulberg is an avid bird watcher, and several tracks by him prominently feature sampled bird sounds and even are titled after his favourite specimens.The productions of The Jewelled Antler Collective often use field recordings featuring birdsong.

In 2007, The CT Collective issed two free albums devoted to music made using bird songs (one with human interaction, one without). The project was co-ordinated by looping musician Nick Robinson

Bird Song and Poetry

Bird song is a popular subject in poetry. Famous poems inspired by bird song include Percy Bysshe Shelley's To a Skylark ("Hail to thee, blithe Spirit!/Bird thou never wert") and Gerard Manley Hopkins' Sea and Skylark. Birdsongs and their relations to Middle-earth inhabitants are a common motif in J. R. R. Tolkien's literary work. The Grateful Dead performed a song called "Bird Song" that Jerry Garcia wrote and dedicated to Janis Joplin.

Bird Intelligence

Figure: *Kea are known for their intelligence and curiosity, both vital traits for survival in the harsh mountain environment that is their home. Kea can solve logical puzzles, such as pushing and pulling things in a certain order to get to food, and will work together to achieve a certain objective. Some have the intelligence of an average six-year-old human.*

Bird intelligence deals with the definition of intelligence and its measurement as it applies to birds. Traditionally, birds have been considered inferior in intelligence to mammals, and derogatory terms such as *bird brains* have been used colloquially in some cultures. Such perceptions are no longer considered scientifically valid. The difficulty of defining or measuring intelligence in non-human animals makes the subject difficult for scientific study. Anatomically, a bird has a relatively large brain compared to its head size. The visual and auditory senses are well developed in most species, while the tactile and olfactory senses are well realised only in a few groups. Birds achieve locomotion through flight and use of the legs in most species. The beak and feet are used to manipulate food and other objects. Birds can communicate using visual signals as well as through the use of calls and song. The testing of intelligence is therefore based on studying the responses to sensory stimuli.

Studies

Bird intelligence has been studied through several attributes and abilities. Many of these studies have been on birds such as quail, domestic fowl and pigeons kept under captive conditions. It has, however, been noted that field studies have been limited, unlike those of the apes. Birds in the crow family (corvids), and parrots (psittacines) have been shown to live socially, have long developmental periods, and possess large forebrains, and these may be expected to allow for greater cognitive abilities.

Counting

Counting has been considered an ability that shows intelligence. Anecdotal evidence from the '60s has suggested that crows may count up to 3. Researchers however need to be cautious and ensure that birds are not merely demonstrating the ability to subitize, or count a small number of items quickly. Some studies, such as one done at Moscow State University, have suggested that crows may indeed have a true numerical ability. It has been shown that parrots can count up to 6.

Cormorants used by Chinese fishermen that were given every eighth fish as a reward were found to be able to keep count up to eight. E.H. Hoh wrote in *Natural History* magazine:

In the 1970s, on the Li River, Pamela Egremont observed fishermen who allowed the birds to eat every eighth fish they caught. Writing in the *Biological Journal of the Linnean Society,* she reported that, once their quota of seven fish was filled, the birds "stubbornly refuse to move again until their neck ring is loosened. They ignore an order to dive and even resist a rough push or a knock, sitting glum and motionless on their perches." Meanwhile, other birds that had not filled their quotas continued to catch fish as usual. "One is forced to conclude that these highly intelligent birds can count up to seven," she wrote.

Many birds are also able to detect changes in the number of eggs in their nest and brood. Parasitic cuckoos are often known to remove one of the host eggs before laying their own.

Associative Learning

Visual or auditory signals and their association with food and other rewards have been well studied, and birds have been trained to recognise and distinguish complex shapes. This is probably an important ability that aids their survival.

Spatial and Temporal Abilities

A common test of intelligence is the detour test where a glass barrier between the bird and an item such as food is used in the setup. Most mammals discover that the objective is reached by first going away from the target. Whereas domestic fowl fail on this test, many within the crow family are able to readily solve the problem.

Large fruit-eating birds in tropical forests depend on trees which bear fruit at different times of the year. Many species, such as pigeons and hornbills, have been shown to be able to decide upon foraging areas according to the time of the year. Birds that show food hoarding behaviour have also shown the ability to recollect the locations of food caches. Nectarivorous birds such as hummingbirds also optimise their foraging by keeping track of the locations of good and bad flowers. Studies of Western Scrub Jays also suggests that birds may be able to plan ahead. They cache food according to future needs and risk of not being able to find the food on subsequent days.

Many birds follow strict time schedules in their activities. These are often dependent upon environmental cues. Birds also are sensitive to daylight length, and this awareness is especially important as a cue for migratory species. The ability to orient themselves during migrations is attributed to birds' superior sensory abilities, rather than to intelligence.

Self Awareness

The mirror test allows scientists to check if birds are conscious of themselves and able to distinguish themselves and from other animals by determining whether an animal possesses or lacks the ability to recognise itself in its own reflection. The European Magpie is the only animal, besides mammals that is shown to be able to pass this test by trying to remove a coloured sticker from underneath their beaks when shown in a mirror.

However in 1981, Epstein, Lanza and Skinner published a paper in the journal *Science* in which they argued that the pigeon also passes the mirror test. A pigeon was trained to look in a mirror to find a response key behind it which it then turned to peck - food was the consequence of a correct choice (i.e., the pigeon learned to use a mirror to find critical elements of its environment). Next, the bird was trained to peck at dots placed on its feathers; food was, again, the consequence of touching the dot.

The latter training was accomplished in the absence of the mirror. The final test was placing a small bib on the pigeon - enough to cover a dot placed on its lower belly. A control period without the mirror present yielded no pecking at the dot. When the mirror was revealed, the pigeon became active, looked in the mirror and then tried to peck on the dot under the bib.

It is true that untrained pigeons have never been able to pass the mirror test. However, pigeons do not normally have access to mirrors and do not have the necessary experiences to use them. Giving the pigeons this experience in no way guaranteed it would pass the mirror test: Remember, the pigeon never pecked dots on its own body in the presence of the mirror (until the final test).

Despite that the birds are not classified as being able to recognise their reflection, because those that did were trained to do so and the animal has to be able to do this without human assistance. Therefore it must also be shown that the birds are able to do this in the wild as well with no experience, just on their own intelligence to see if it is able to comprehend that it is looking at its own reflection on its own. However even when an animal is trained to do this it's still unknown if they are aware that they are looking at themselves, or are just repeating the same movements and commands that they were taught so that they may receive a treat as a reward after they have correctly completed their task.

Tool Use

Many birds have been shown capable of using tools. The definition of a tool has been debated. One proposed definition of tool use has been defined by T. B. Jones and A. C. Kamil in 1973 as the use of physical objects other than the animal's own body or appendages as a means to extend the physical influence realised by the animal

By this definition, a Bearded Vulture (Lammergeier) dropping a bone on a rock would not be using a tool since the rock cannot be seen as an extension of the body. However the use of a rock manipulated using the beak to crack an ostrich egg would qualify the Egyptian Vulture as a tool user. Many other species, including parrots, corvids and a range of passerines, have been noted as tool users.

New Caledonian Crows have been observed in the wild to use sticks with their beaks to extract insects from logs. While young birds in the wild normally learn this technique from elders, a laboratory crow named "Betty" improvised a hooked tool from a wire with no prior experience.

The Woodpecker Finch from the Galapagos Islands also uses simple stick tools to assist it in obtaining food. In captivity, a young Cactus Finch learned to imitate this behaviour by watching a woodpecker finch in an adjacent cage.

Crows in urban Japan have innovated a technique to crack hard-shelled nuts by dropping them onto crosswalks and letting them be run over and cracked by cars. They then retrieve the cracked nuts when the cars are stopped at the red light. Macaws have been shown to utilise rope to fetch items that would normally be difficult to reach. Striated Herons (*Butorides striatus*) use bait to catch fish.

Observational Learning

Using rewards to reinforce responses is often used in laboratories to test intelligence. However, the ability of animals to learn by observation and imitation is considered more significant. Crows have been noted for their ability to learn from each other.

Brain Anatomy

At the beginning of the 20th century, scientists argued that the birds had hyper-developed basal ganglia, with tiny mammalian-like telencephalon structures. Modern studies have refuted this view. The basal ganglia only occupy a small part of the avian brain. Instead, it seems that birds use a different part of their brain, the medio-rostral neostriatum/hyperstriatum ventrale, as the seat of their intelligence, and the brain-to-body size ratio of psittacines (parrots) and corvines (birds of the crow family) is actually comparable to that of higher primates.

Studies with captive birds have given insight into which birds are the most intelligent. While parrots have the distinction of being able to mimic human speech, studies with the African Grey Parrot have shown that some are able to associate words with their meanings and form simple sentences. Parrots and the corvid family of crows, ravens, and jays are considered the most intelligent of birds. Not surprisingly, research has shown that these species tend to have the largest HVCs. Dr. Harvey J. Karten, a neuroscientist at UCSD who has studied the physiology of birds, has discovered that the lower parts of avian brains are similar to those of humans.

Social Behaviour

Social life has been considered to be a driving force for the evolution of intelligence. Many birds have social organisations, and loose aggregations are common. Many corvid species separate into small family groups (or "clans") for activities such as nesting and territorial defence. The birds then congregate in massive flocks made up of several different species for migratory purposes. Some birds use teamwork while hunting. Predatory birds hunting in pairs have been observed using a "bait and switch" technique, whereby one bird will distract the prey while the other swoops in for the kill.

Social behaviour requires individual identification, and most birds appear to be capable of recognising mates, siblings and young. Other behaviours such as play and cooperative breeding are also considered indicators of intelligence. When crows are caching food, they appear to be sensitive to note who is watching them hide the food. They also steal food caught by others.

In some fairy-wrens such as the Superb and Red-backed, males pick flower petals in colours contrasting with their bright nuptial plumage and present them to others of their species that will acknowledge, inspect and sometimes manipulate the petals. This function seems not linked to sexual or aggressive activity in the short and medium term thereafter, though its function is apparently not aggressive and quite possibly sexual.

Language

Birds communicate with their flockmates through song, calls, and body language. Studies have shown that the intricate territorial songs of some birds must be learned at an early age, and that the memory of the song will serve the bird for the rest of its life. Some bird species are able to communicate in a variety of dialects. For example, the New Zealand saddleback will learn the different song "dialects" of clans of its own species, much as human beings might learn diverse regional dialects. When a territory-owning male of the species dies, a young male will immediately take his place, singing to prospective mates in the dialect appropriate to the territory he is in.

Recent studies indicate that some birds may have an ability to understand grammatical structures.

Conceptual Abilities

Evidence that birds can form abstract concepts such as *same v. different* has been proven by Alex, the African Grey Parrot. Alex was trained by animal psychologist Irene Pepperberg to vocally label more than 100 objects of different colours and shapes and which are made from different materials. Alex could also request or refuse these objects ('I want X') and quantify numbers of them.

It has been noted that bird brainwaves are very similar to that of humans, there are many similarities between the lobes of the brain and most notable the cerebral cortex. For this reason it is believed they can understand human language.

Object Permanence

Macaws have been demonstrated to fully comprehend the concept of object permanence at a young age. They will even refute the "A-not-B error". If they are shown an item, especially one with whose purpose they are familiar - they will search logically for where it could be feasibly placed. One test for YouTube was done as follows: A macaw was shown an item; the item was then hidden behind the back of the trainer and placed into a container. The container it was placed in without the macaw seeing, along with another container and multiple objects, were spread upon a table simultaneously. The specific container that the item was stored in out of the macaws' sight was one that the macaw had never observed before. The macaw searched this some, then another container, then returning to open the correct container to demonstrate knowledge of and the ability to search for the item.

Theory of Mind

A study on the Little Green Bee-eater suggests that these birds may be able to see from the point of view of a predator. The Western Scrub Jay hides caches of food and will later re-hide food if it was watched by another bird the first time, but only if the bird hiding the food has itself stolen food before from a cache. This might suggest a theory of mind, but other "lower level" explanations are possible. Such an ability to see from the point of view of another individual had previously been attributed only to the great apes. Such abilities form the basis for empathy. Research published in 2008 that was conducted with an Eleonora Cockatoo named Snowball has shown that birds can learn to dance to man-made music.

Bird Migration

Bird migration is the regular seasonal journey undertaken by many species of birds. Bird movements include those made in response to changes in food availability, habitat or weather. Sometimes, journeys are not termed "true migration" because they are irregular (nomadism, invasions, irruptions) or in only one direction (dispersal, movement of young away from natal area). Migration is marked by its annual seasonality. In contrast, birds that are non-migratory are said to be resident or sedentary. Approximately 1800 of the world's 10,000 bird species are long-distance migrants.

Figure 6: *A flock of Barnacle Geese during autumn migration*

General Patterns

Many bird populations migrate long distances along a flyway. The most common pattern involves flying north in the spring to breed in the temperate or Arctic summer and returning in the autumn to wintering grounds in warmer regions to the south. Of course, in the Southern Hemisphere the directions are reversed, but there is less land area in the far South to support long-distance migration.

The primary motivation for migration appears to be food; for example, some hummingbirds choose not to migrate if fed through the winter. Also, the longer days of the northern summer provide extended time for breeding birds to feed their young. This helps diurnal birds to produce larger clutches than related non-migratory species that remain in the tropics. As the days shorten in autumn, the birds return to warmer regions where the available food supply varies little with the season. These advantages offset the high stress, physical exertion costs, and other risks of the migration such as predation. Predation can be heightened during migration: the Eleonora's Falcon, which breeds on Mediterranean islands, has a very late breeding season, coordinated with the autumn passage of southbound passerine migrants, which it feeds to its young. A similar strategy is adopted by the Greater Noctule bat, which preys on nocturnal passerine migrants. The higher concentrations of migrating birds at stopover sites make them prone to parasites and pathogens, which require a heightened immune response.

Figure 7: *Migrating waders in Roebuck Bay, Western Australia*

Within a species not all populations may be migratory; this is known as "partial migration". Partial migration is very common in the southern continents; in Australia, 44% of non-passerine birds and 32% of passerine species are partially migratory. In some species, the population at higher latitudes tends to be migratory and will often winter at lower latitude. The migrating birds bypass the latitudes where other populations may be sedentary, where suitable wintering habitats may already be occupied.

This is an example of *leap-frog migration*. Many fully migratory species show leap-frog migration (birds that nest at higher latitudes spend the winter at lower latitudes), and many show the alternative, "chain migration" where populations 'slide' more evenly North and South without reversing order.

Within a population, it is common for different ages and/or sexes to have different patterns of timing and distance. Only the female Chaffinches in Scandinavia migrate, with the males staying resident. This has given rise to the latter's specific name of *coelebs*, a bachelor.

Most migrations begin with the birds starting off in a broad front. Often, this front narrows into one or more preferred routes termed flyways. These routes typically follow mountain ranges or coastlines, sometimes rivers, and may take advantage of updrafts and other wind patterns or avoid geographical barriers such as large stretches of open water. The specific routes may be genetically programmed or learned to varying degrees. The routes taken on forward and return migration are often different. A common pattern in North America is clockwise migration, where birds flying North tend to be further West, and flying South tend to shift Eastwards.

Many, if not most, birds migrate in flocks. For larger birds, flying in flocks reduces the energy cost. Geese in a V-formation may conserve 12–20 % of the energy they would need to fly alone. Red Knots *Calidris canutus* and Dunlins *Calidris alpina* were found in radar studies to fly 5 km per hour faster in flocks than when they were flying alone.

Birds fly at varying altitudes during migration. An expedition to Mt. Everest found skeletons of Pintail and Black-tailed Godwit at 5000 m (16,400 ft) on the Khumbu Glacier. Bar-headed Geese have been seen flying over the highest peaks of the Himalayas above 8000 m (29000 ft) even when low passes of 3000 m (10000 ft) were nearby. Seabirds fly low over water but gain altitude when crossing land, and the reverse pattern is seen in landbirds. However most bird migration is in the range of 150 m (500 ft) to 600 m (2000 ft). Bird-hit aviation records from the United States show most collisions occur below 600 m (2000 ft) and almost none above 1800 m (6000 ft).

Bird migration is not limited to birds that can fly. Most species of penguin migrate by swimming. These routes can cover over 1000 km. Blue Grouse *Dendragapus obscurus* perform altitudinal migration mostly by walking. Emus in Australia have been observed to undertake long-distance movements on foot during droughts.

Historical Views

Records of bird migration were made 3000 years ago by Hesiod, Homer, Herodotus and Aristotle. The Bible also notes migrations, as in the Book of Job (39:26), where the inquiry is made: "Doth the hawk fly by Thy wisdom and stretch her wings toward the south?" The author of Jeremiah (8:7) wrote: "The stork in the heavens knoweth her appointed time; and the turtledove, and the crane, and the swallow, observe the time of their coming."

Aristotle noted that cranes travelled from the steppes of Scythia to marshes at the headwaters of the Nile. Pliny the Elder, in his *Historia*

Naturalis, repeats Aristotle's observations. Aristotle however suggested that swallows and other birds hibernated. This belief persisted as late as 1878, when Elliott Coues listed the titles of no less than 182 papers dealing with the hibernation of swallows. It was not until early in the nineteenth century that migration as an explanation for the winter disappearance of birds from northern climes was accepted.

Long-distance Migration

The typical image of migration is of northern landbirds, such as swallows and birds of prey, making long flights to the tropics. Many northern-breeding ducks, geese and swans are also long-distance migrants, but need only to move from their Arctic breeding grounds far enough south to escape frozen waters. Most Holarctic wildfowl species remain in the Northern Hemisphere, but in countries with milder climates.

For example, the pink-footed goose migrates from Iceland to Britain and neighbouring countries. Migratory routes and wintering grounds are traditional and learned by young during their first migration with their parents. Some ducks, such as the Garganey, move completely or partially into the tropics.

The same considerations about barriers and detours that apply to long-distance land-bird migration apply to water birds, but in reverse: a large area of land without bodies of water that offer feeding sites is a barrier to a water bird.

Open sea may also be a barrier to a bird that feeds in coastal waters. Detours avoiding such barriers are observed: for example, Brent Geese migrating from the Taymyr Peninsula to the Wadden Sea travel via the White Sea coast and the Baltic Sea rather than directly across the Arctic Ocean and northern Scandinavia.

Figure: *Bar-tailed Godwit*

A similar situation occurs with waders (called "shorebirds" in North America). Many species, such as Dunlin and Western Sandpiper, undertake long movements from their Arctic breeding grounds to warmer locations in the same hemisphere, but others such as Semipalmated Sandpiper travel longer distances to the tropics in the Southern Hemisphere. Like the large and powerful wildfowl, the waders are strong fliers. This means that birds wintering in temperate regions have the capacity to make further shorter movements in the event of particularly inclement weather.

For some species of waders, migration success depends on the availability of certain key food resources at stopover points along the migration route. This gives the migrants an opportunity to "refuel" for the next leg of the voyage. Some examples of important stopover locations are the Bay of Fundy and Delaware Bay.

Some Bar-tailed Godwits have the longest known non-stop flight of any migrant, flying 11,000 km from Alaska to their New Zealand non-breeding areas. Prior to migration, 55 percent of their bodyweight is stored fat to fuel this uninterrupted journey.

Figure: *Arctic Terns*

Seabird migration is similar in pattern to those of the waders and waterfowl. Some, such as the Black Guillemot and some gulls, are quite sedentary; others, such as most terns and auks breeding in the temperate northern hemisphere, move varying distances south in the northern winter.

The Arctic Tern has the longest-distance migration of any bird, and sees more daylight than any other, moving from its Arctic breeding grounds to the Antarctic non-breeding areas. One Arctic Tern, ringed (banded) as a chick on the Farne Islands off the British east coast, reached Melbourne, Australia in just three months from fledging, a sea journey of over 22,000 km (14,000 mi). A few seabirds, such as Wilson's Petrel and Great Shearwater, breed in the southern hemisphere and migrate north in the southern winter. Seabirds have the additional advantage of being able to feed during migration over open waters.

The most pelagic species, mainly in the 'tubenose' order Procellariiformes, are great wanderers, and the albatrosses of the southern oceans may circle the globe as they ride the "roaring forties" outside the breeding season. The tubenoses spread widely over large areas of open ocean, but congregate when food becomes available. Many are also among the longest-distance migrants; Sooty Shearwaters nesting on the Falkland Islands migrate 14,000 km (8,700 mi) between the breeding colony and the North Atlantic Ocean off Norway. Some Manx Shearwaters do this same journey in reverse. As they are long-lived birds, they may cover enormous distances during their lives; one record-breaking Manx Shearwater is calculated to have flown 8 million km (5 million miles) during its over-50 year lifespan.

Figure: *Griffon Vulture soaring*

Some large broad-winged birds rely on thermal columns of rising hot air to enable them to soar. These include many birds of prey such as vultures, eagles, and buzzards, but also storks. These birds migrate in the daytime. Migratory species in these groups have great difficulty crossing large bodies of water, since thermals only form over land, and these birds cannot maintain active flight for long distances.

The Mediterranean and other seas present a major obstacle to soaring birds, which must cross at the narrowest points. Massive numbers of large raptors and storks pass through areas such as

Gibraltar, Falsterbo, and the Bosphorus at migration times. More common species, such as the Honey Buzzard, can be counted in hundreds of thousands in autumn. Other barriers, such as mountain ranges, can also cause funnelling, particularly of large diurnal migrants. This is a notable factor in the Central American migratory bottleneck.

Many of the smaller insectivorous birds including the warblers, hummingbirds and flycatchers migrate large distances, usually at night. They land in the morning and may feed for a few days before resuming their migration. The birds are referred to as *passage migrants* in the regions where they occur for short durations between the origin and destination. Nocturnal migrants minimise predation, avoid overheating, and feed during the day. One cost of nocturnal migration is the loss of sleep. Migrants may be able to alter their quality of sleep to compensate for the loss.

Short-distance and Altitudinal Migration

Many long-distance migrants appear to be genetically programmed to respond to changing day length. Species that move short distances, however, may not need such a timing mechanism, and may move in response to local weather conditions.

Figure: *Ruby-throated Hummingbird*

Thus mountain and moorland breeders, such as Wallcreeper and White-throated Dipper, may move only altitudinally to escape the cold higher ground. Other species such as Merlin and Skylark will move further to the coast or to a more southerly region. Species like the Chaffinch are not migratory in Britain, but will move south or to Ireland in very cold weather. Short-distance passerine migrants have two evolutionary origins. Those that have long-distance migrants in the same family, such as the Chiffchaff, are species of southern hemisphere origins that have progressively shortened their return migration to stay in the northern hemisphere.

Species that have no long-distance migratory relatives, such as the waxwings, are effectively moving in response to winter weather, rather than enhanced breeding opportunities.

In the tropics there is little variation in the length of day throughout the year, and it is always warm enough for a food supply (although because of competition, there may not be enough food for every bird). Migration within the tropics has been far less studied than in the temperate zones. It was once assumed that tropical birds were mostly sedentary; however, altitudinal migration and other within-tropics movements appear to be surprisingly common. Many tropical regions have wet and dry seasons, inducing some birds to migrate or wander widely to find food. Indeed, the monsoons of India are preceded by the arrival of the Jacobin Cuckoo, the "harbinger of the monsoon". Other examples include the Woodland Kingfisher of west Africa and many Australian birds.

Figure: *Rainbow Bee-eater*

There are a few species, notably cuckoos, which are genuine long-distance migrants within the tropics. An example is the Lesser Cuckoo, which breeds in India and spends the non-breeding season in Africa. Such examples help make the case that food supplies, not weather per se, drive migration patterns.

Altitudinal migration is common on mountains worldwide, such as in the Himalayas and the Andes. Quite often, altitudinal migration is combined with distance migration; for example, the Himalayan

Kashmir Flycatcher and Pied Thrush both move as far south as the highlands of Sri Lanka. Altitudinal migration may even be important to birds living on relatively small islands, such as the Hawaiian Islands, which have high mountains.

Irruptions and Dispersal

Sometimes circumstances such as a good breeding season followed by a food source failure the following year lead to irruptions in which large numbers of a species move far beyond the normal range. Bohemian Waxwing and Common Crossbills show this unpredictable variation in annual numbers. The temperate zones of the southern continents have extensive arid areas, particularly in Australia and western southern Africa, and weather-driven movements are common but not always predictable. A couple of weeks of heavy rain in one part or another of the usually dry centre of Australia, for example, causes dramatic plant and invertebrate growth, attracting birds from all directions. This can happen at any time of year, and, in any given area, may not happen again for a decade or more, depending on the frequency of El Niño and La Niña periods.

Bird migration is primarily, but not entirely, a Northern Hemisphere phenomenon. In the Southern Hemisphere, seasonal migration tends to be much less obvious. There are several reasons for this. First, the largely uninterrupted expanses of land mass or ocean tend not to funnel migrations into narrow and obvious pathways, making them less obvious to the human observer. Second, at least for terrestrial birds, climatic regions tend to fade into one another over a long distance rather than be entirely separate: this means that rather than make long trips over unsuitable habitat to reach particular destinations, migrant species can usually travel at a relaxed pace, feeding as they go. Short of banding studies it is often not obvious that the birds seen in any particular locality as the seasons change are in fact different members of the same species passing through, gradually working their way north or south.

Many species do in fact breed in the temperate southern hemisphere regions and winter further north in the tropics. The southern African Greater Striped Swallow, and the Australian Satin Flycatcher, Dollarbird, and Rainbow Bee-eater for example, winters well north of their breeding range.

Physiology and Control

The control of migration, its timing and response are genetically controlled and appear to be a primitive trait that is present even in non-migratory species of birds. The ability to navigate and orient themselves during migration is a much more complex phenomenon that may include both endogenous programs as well as learning.

Timing

The primary physiological cue for migration are the changes in the day length. These changes are also related to hormonal changes in the birds.

In the period before migration, many birds display higher activity or Zugunruhe (German: *migratory restlessness*) as well as physiological changes such as increased fat deposition.

The occurrence of Zugunruhe even in cage-raised birds with no environmental cues (e.g. shortening of day and falling temperature) has pointed to the role of circannual endogenous programs in controlling bird migrations. Caged birds display a preferential flight direction that corresponds with the migratory direction they would take in nature, even changing their preferential direction at roughly the same time their wild conspecifics change course.

In species where there is polygyny and with considerable sexual dimorphism, there is a tendency for males to return earlier to the breeding sites than their females. This is termed as protandry.

Orientation and Navigation

Navigation is based on a variety of senses. Many birds have been shown to use a sun compass. Using the sun for direction involves the need for making compensation based on the time. Navigation has also been shown to be based on a combination of other abilities including the ability to detect magnetic fields (magnetoception), use visual landmarks as well as olfactory cues.

Long distance migrants are believed to disperse as young birds and form attachments to potential breeding sites and to favourite wintering sites. Once the site attachment is made they show high site-fidelity, visiting the same wintering sites year after year.

The ability of birds to navigate during migrations cannot be fully explained by endogenous programming, even with the help of responses to environmental cues. The ability to successfully perform long-distance migrations can probably only be fully explained with an accounting for the cognitive ability of the birds to recognise habitats and form mental maps. Satellite tracking of day migrating raptors such as Ospreys and Honey Buzzards has shown that older individuals are better at making corrections for wind drift.

As the circannual patterns indicate, there is a strong genetic component to migration in terms of timing and route, but this may be modified by environmental influences. An interesting example where a change of migration route has occurred because of such a geographical barrier is the trend for some Blackcaps in central Europe to migrate west and winter in Britain rather than cross the Alps.

Migratory birds may use two electromagnetic tools to find their destinations: one that is entirely innate and another that relies on experience. A young bird on its first migration flies in the correct direction according to the Earth's magnetic field, but does not know how far the journey will be.

It does this through a radical pair mechanism whereby chemical reactions in special photo pigments sensitive to long wavelengths are affected by the field. Note that although this only works during daylight hours, it does not use the position of the sun in any way. At this stage the bird is similar to a boy scout with a compass but no map, until it grows accustomed to the journey and can put its other facilities to use.

With experience they learn various landmarks and this "mapping" is done by magnetites in the trigeminal system, which tell the bird how strong the field is. Because birds migrate between northern and southern regions, the magnetic field strengths at different latitudes let it interpret the radical pair mechanism more accurately and let it know when it has reached its destination.

More recent research has found a neural connection between the eye and "Cluster N", the part of the forebrain that is active during migrational orientation, suggesting that birds may actually be able to *see* the magnetic field of the earth.

Vagrancy

Migrating birds can lose their way and appear outside their normal ranges. This can be due to flying past their destinations as in the "spring overshoot" in which birds returning to their breeding areas overshoot and end up further north than intended. Reverse migration, where the genetic programming of young birds fails to work properly, can lead to great rarities turning up as vagrants thousands of kilometres out of range.

Certain areas, because of their location, have become famous as watchpoints for migrating birds. Examples are the Point Pelee National Park in Canada, and Spurn in England. Drift migration of birds blown off course by the wind can result in "falls" of large numbers of migrants at coastal sites.

Migration Conditioning

It has been possible to teach a migration route to a flock of birds, for example in reintroduction schemes. After a trial with Canada Geese, microlight aircraft were used in the US to teach safe migration routes to reintroduced Whooping Cranes.

Adaptations

Birds need to alter their metabolism in order to meet the demands of migration. The storage of energy through the accumulation of fat and the control of sleep in nocturnal migrants require special physiological adaptations. In addition, the feathers of a bird suffer from wear-and-tear and require to be molted.

The timing of this molt - usually once a year but sometimes two - varies with some species molting prior to moving to their winter grounds and others molting prior to returning to their breeding grounds. Apart from physiological adaptations, migration sometimes requires behavioural changes such as flying in flocks to reduce the energy used in migration or the risk of predation.

Evolutionary and Ecological Factors

Migration in birds is highly labile and is believed to have developed independently in many avian lineages. While it is agreed that the behavioural and physiological adaptations necessary for migration are under genetic control, some authors have argued that no genetic change is necessary for migratory behaviour to develop in a sedentary species because the genetic framework for migratory behaviour exists in nearly all avian lineages. This explains the rapid appearance of migratory behaviour after the most recent glacial maximum.

Whether a particular species migrates depends on a number of factors. The climate of the breeding area is important, and few species can cope with the harsh winters of inland Canada or northern Eurasia. Thus the partially migratory Blackbird *Turdus merula* is migratory in Scandinavia, but not in the milder climate of southern Europe. The nature of the staple food is also significant. Most specialist insect eaters outside the tropics are long-distance migrants, and have little choice but to head south in winter.

Sometimes the factors are finely balanced. The Whinchat *Saxicola rubetra* of Europe and the Siberian Stonechat *Saxicola maura* of Asia are long-distance migrants wintering in the tropics, whereas their close relative, the European Stonechat *Saxicola rubicola* is a resident bird in most of its range, and moves only short distances from the colder north and east. A possible factor here is that the resident species can often raise an extra brood. Recent research suggests that long-distance passerine migrants are of South American and African, rather than northern hemisphere, evolutionary origins. They are effectively southern species coming north to breed rather than northern species going south to winter.

Theoretical analyses, summarised by Alerstam (2001), show that detours that increase flight distance by up to 20% will often be adaptive on aerodynamic grounds - a bird that loads itself with food to cross a long barrier flies less efficiently. However some species show circuitous migratory routes that reflect historical range expansions and are far from optimal in ecological terms. An example is the migration of continental populations of Swainson's Thrush, which fly far east across North America before turning south via Florida to reach northern South America; this route is believed to be the consequence of a range expansion that occurred about 10,000 years ago. Detours may also be caused by differential wind conditions, predation risk, or other factors.

Climate Change

Large scale climatic changes, as have been experienced in the past, are expected to have an effect on the timing of migration. Studies have shown a variety of effects including timing changes in migration, breeding as well as population variations.

Ecological Effects

The migration of birds also aids the movement of other species, including those of ectoparasites such as ticks and lice, which in turn may carry micro-organisms including those of concern to human health. Considerable interest has been taken due to the global spread of avian influenza, however migrant birds have not been found to be a special risk, with import of pet and domestic birds being a greater threat. Some viruses that are maintained in birds without lethal effects, such as the West Nile Virus may however be spread by migrating birds. Birds may also have a role in the dispersal of propagules of plants and plankton. Some predators take advantage of the concentration of birds during migration. Greater Noctule bats feed on nocturnal migrating passerines. Some birds of prey specialise on migrating waders.

Study Techniques

Early studies on the timing of migration began in 1749 in Finland, with Johannes Leche of Turku collecting the dates of arrivals of spring migrants.

Bird migration routes have been studied by a variety of techniques of which ringing is the oldest. Colour marking and use of radar, satellite tracking are some of the other techniques. Stable isotopes of hydrogen, oxygen, carbon, nitrogen, and sulphur have also been used to establish avian migratory connectivity between wintering sites and breeding grounds. Stable isotopic methods to establish migratory linkage rely on spatial isotopic differences in bird diet that are incorporated into inert tissues like feathers, or into growing tissues such as claws and muscle or blood. An approach to identify migration intensity makes use of upward pointing microphones to record the nocturnal contact calls of flocks flying overhead. These are then analysed in a laboratory to measure time, frequency and species.

Figure: *Emlen funnel*

An older technique to quantify migration involves observing the face of the moon towards full moon and counting the silhouettes of flocks of birds as they fly at night. Orientation behaviour studies have been traditionally carried out using variants of a setup known as the Emlen funnel, which consists of a circular cage with the top covered by glass or wire-screen so that either the sky is visible or the setup is placed in a planetarium or with other controls on environmental cues.

The orientation behaviour of the bird inside the cage is studied quantitatively using the distribution of marks that the bird leaves on the walls of the cage. Other approaches used in pigeon homing studies make use of the direction in which the bird vanishes on the horizon.

Threats and Conservation

Human activities have threatened many migratory bird species. The distances involved in bird migration mean that they often cross political boundaries of countries and conservation measures require international cooperation. Several international treaties have been signed to protect migratory species including the Migratory Bird Treaty Act of 1918 of the US and the African-Eurasian Migratory Waterbird Agreement.

The concentration of birds during migration can put species at risk. Some spectacular migrants have already gone extinct, the most notable being the Passenger Pigeon (*Ectopistes migratorius*). During migration the flocks were a mile (1.6 km) wide and 300 miles (480 km) long, taking several days to pass and containing up to a billion birds.

Other significant areas include stop-over sites between the wintering and breeding territories. A capture-recapture study of passerine migrants with high fidelity for breeding and wintering sites did not show similar strict association with stop-over sites.

Hunting along the migratory route can also take a heavy toll. The populations of Siberian Cranes that wintered in India declined due to hunting along the route, particularly in Afghanistan and Central Asia. Birds were last seen in their favourite wintering grounds in Keoladeo National Park in 2002.

Structures such as power lines, wind farms and offshore oil-rigs have also been known to affect migratory birds. Habitat destruction by land use changes is the biggest threat, and shallow wetlands that are stopover and wintering sites for migratory birds are particularly threatened by draining and reclamation for human use.

Chapter 6

Behaviour of Cats

Cat Anatomy

Mouth

Cats have highly specialised teeth for the killing of prey and the tearing of meat. The *premolar* and *first molar*, together the *carnassial pair* are located on each side of the mouth. These teeth efficiently function to shear meat like a pair of scissors.

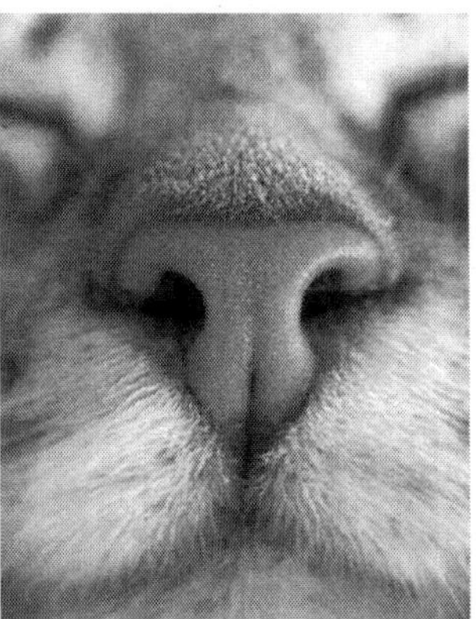

Figure: *A cat's nose is highly adapted.*

While this is present in canids, it is highly developed in felines. The cat's tongue has sharp spines, or *papillae*, useful for retaining and ripping flesh from a carcass. These papillae are small backward-facing hooks that contain keratin which also assist in their groom. The cat's oral structures provide for a variety of vocalisations used for communication, including meowing, purring, hissing, growling, squeaking, chirping, clicking, and grunting. Cats also employ a variety of body language: position of ears and tail, relaxation of whole body, kneading of paws, all are indicators of mood.

Ears

Thirty-two individual muscles in each ear allow for a manner of directional hearing; a cat can move each ear independently of the other. Because of this mobility, a cat can move its body in one direction and point its ears in another direction. Most cats have straight ears pointing upward. Unlike dogs, flap-eared breeds are extremely rare (*Scottish Folds* are one such exceptional mutation). When angry or frightened, a cat will lay back its ears to accompany the growling or hissing sounds it makes. Cats also turn their ears back when they are playing or to listen to a sound coming from behind them. The angle of cats' ears is an important clue to their mood.

Nose

The cats are highly territorial and secreting odors plays a major role in cat communication. The nose helps cats to identify territories, other cats and mates, to locate food, and for various other causes. A cat's sense of smell is believed to be about fourteen times stronger than that of humans. The bit of nose we see, the nose leather is quite tough to allow it to absorb rather rough treatment sometimes. The colour varies according to the genotype (genetic makeup) of the cat. Cat's skin has the same colour as the fur but the colour of the nose leather is probably dictated by a dedicated gene. Cats with white fur have skin susceptible to damage by ultraviolet light that may cause cancer. Extra care is required when she/he goes outside in hot sun.

Legs

Cats, like dogs, are digitigrades. They walk directly on their toes, with the bones of their feet making up the lower part of the visible leg. Cats are capable of walking very precisely because like all felines, they directly register; that is, they place each hind paw (almost) directly in the print of the corresponding forepaw, minimising noise and visible tracks. This also provides sure footing for their hind paws when they navigate rough terrain. The two back legs make the cat able to leap far distances and fall from high places without getting hurt.

Claws

Like nearly all members of the family Felidae, cats have retractable claws. In their normal, relaxed position, the claws are sheathed with the skin and fur around the toe pads. This keeps the claws sharp by preventing wear from contact with the ground and allows the silent stalking of prey. The claws on the forefeet are typically sharper than those on the hind feet. Cats can voluntarily extend their claws on one or more paws.

They may extend their claws in hunting or self-defense, climbing, "kneading", or for extra traction on soft surfaces (bedspreads, thick rugs, skin, etc.). It is also possible to make a cooperative cat extend its claws by carefully pressing both the top and bottom of the paw. The curved claws can become entangled in carpet or thick fabric, which can cause injury if the cat is unable to free itself.

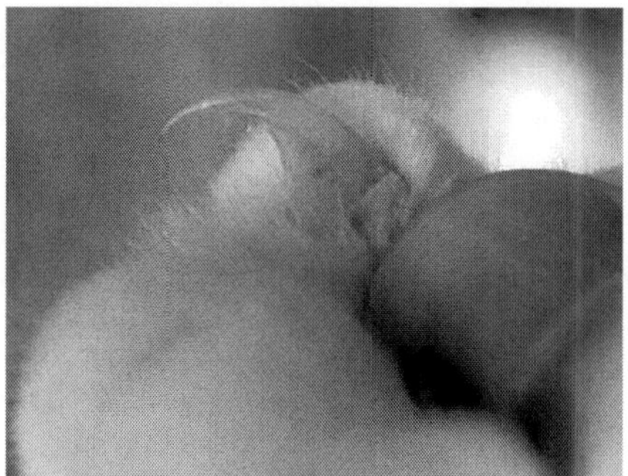

Figure: *A cat's claw*

Most cats have five claws on their front paws, and four or five on their rear paws. Because of an ancient mutation, however, domestic and feral cats are prone to polydactylyism, (particularly in the east coast of Canada and northeast coast of the United States) and can have six or seven toes. The fifth front claw (the *dewclaw*) is proximal to the other claws. There is a protrusion which appears to be a sixth "finger". This special feature of the front paws, on the inside of the wrists, is the carpal pad, also found on the paws of big cats and dogs. It has no function in normal walking, but is thought to be an anti-skidding device used while jumping.

Temperature and Heart Rate

The normal body temperature of a cat is between 38 and 39 °C (101 and 102.2 °F). A cat is considered *febrile* (hyperthermic) if it has a temperature of 39.5 °C (103 °F) or greater, or *hypothermic* if less than 37.5 °C (100 °F). For comparison, humans have a normal temperature of approximately 36.8 °C (98.6 °F). A domestic cat's normal heart rate ranges from 140 to 220 beats per minute, and is largely dependent on how excited the cat is. For a cat at rest, the average heart rate usually is between 150 and 180 bpm, about twice that of a human (average 80 bpm).

Figure: *Two cats sharing body heat*

Skin

Cats possess rather loose skin; this allows them to turn and confront a predator or another cat in a fight, even when it has a grip on them. This is also an advantage for veterinary purposes, as it simplifies injections.

In fact, the lives of cats with kidney failure can sometimes be extended for years by the regular injection of large volumes of fluid subcutaneously, which serves as an alternative to dialysis.

The particularly loose skin at the back of the neck is known as the *scruff*, and is the area by which a mother cat grips her kittens to carry them. As a result, cats tend to become quiet and passive when gripped there. This behaviour also extends into adulthood, when a male will grab the female by the scruff to immobilise her while he mounts, and to prevent her from running away as the mating process takes place.

This technique can be useful when attempting to treat or move an uncooperative cat. However, since an adult cat is heavier than a kitten, a pet cat should never be carried by the scruff, but should instead have its weight supported at the rump and hind legs, and at the chest and front paws. Often (much like a small child), a cat will lie with its head and front paws over a person's shoulder, and its back legs and rump supported under the person's arm.

Skeleton

Cats have seven cervical vertebrae like almost all mammals, thirteen thoracic vertebrae (humans have twelve), seven lumbar vertebrae (humans have five), three sacral vertebrae (humans have five because of their bipedal posture), and, except for Manx cats, twenty-two or twenty-three caudal vertebrae (humans have three to five, fused into an internal coccyx).

The extra lumbar and thoracic vertebrae account for the cat's enhanced spinal mobility and flexibility, compared to humans. The caudal vertebrae form the *tail*, used by the cat as a counterbalance to the body during quick movements. Cats also have free-floating clavicle bones, which allows them to pass their body through any space into which they can fit their heads.

Muscles

Abdominal:

External Abdominal Oblique: This muscle's origin is the lumbodorsal fascia and ribs. Its insertion is at the pubis and linea alba (via aponeurosis). Its action is to compress abdominal contents. It also laterally flexes and rotates the vertebral column.

Internal Abdominal Oblique: This muscle's origin is the lumbodorsal fascia and pelvis. Its insertion is the linea alba (via aponeurosis), and its action is the compression of abdominal contents. It also laterally flexes and rotates the vertebral column.

Transversus Abdominis

This muscle is the innermost abdominal muscle. Its origin is the second sheet of the lumbodorsal fascia and the pelvic girdle and its insertion is the linea alba. Its action is the compression of the abdomen.

Rectus Abdominis

To see this muscle, first remove the extensive aponeurosis situated on the ventral surface of the cat. Its fibers are extremely longitudinal, on each side of the linea alba. It is also traversed by the inscriptiones tendinae, or what others called *myosepta*.

Deltoid

The deltoid muscles lie just lateral to the trapezius muscles, originating from several fibers spanning the clavicle and scapula, converging to insert at the humerus. Anatomically, there are only two deltoids in the cat, the *acromiodeltoid* and the *spinodeltoid*. However, to conform to human anatomy standards, the clavobrachialis is now also considered a deltoid and is commonly referred to as the *clavodeltoid.*

Acromiodeltoid

The acromiodeltoid is the shortest of the deltoid muscles. It lies lateral to (to the side of) the clavodeltoid, and in a more husky cat it can only be seen by lifting or reflecting the clavodeltoid. It originates at the acromion process and inserts at the deltoid ridge. When contracted, it raises and rotates the humerus outward.

Spinodeltoid

A stout and short muscle lying posterior to the acromiodeltoid. It lies along the lower border of the scapula, and it passes through the upper arm, across the upper end of muscles of the upper arm.

It originates at the spine of the scapula and inserts at the deltoid ridge. Its action is to raise and rotate the humerus outward. Baybaybay/

.

Head

Masseter: The Masseter is a great, powerful, and very thick muscle covered by a tough, shining fascia lying ventral to the zygomatic arch, which is its origin. It inserts into the posterior half of the lateral surface of the mandible. Its action is the elevation of the mandible (closing of the jaw).

Temporalis: The temporalis is a great mass of mandibular muscle, and is also covered by a tough and shiny fascia. It lies dorsal to the zygomatic arch and fills the temporal fossa of the skull. It arises from the side of the skull and inserts into the coronoid process of the mandible. It too, elevates the jaw.

Integumental: The two main integumentary muscles of a cat are the *platysma* and the *cutaneous maximus*. The *cutaneous maximus* covers the dorsal region of the cat and allows it to shake its skin. The *platysma* covers the neck and allows the cat to stretch the skin over the pectoralis major and deltoid muscles.

Neck and Back

Rhomboideus: The Rhomboideus is a thick, large muscle below the Trapezius muscles. It extends from the vertebral border of the scapula to the mid-dorsal line. Its origin is from the neural spines of the first four thoracic vertebrae, and its insertion is at the vertebral border of the scapula. Its action is to draw the scapula to the dorsal.

Rhomboideus Capitis: The Rhomboideus capitis is the most cranial of the deeper muscles. It is underneath the Clavotrapezius. Its origin is the superior nuchal line, and its insertion is at the scapula.

Splenius

The Splenius is the most superficial of all the deep muscles. It is a thin, broad sheet of muscle underneath the Clavotrapezius and deflecting it. It is crossed also by the Rhomboideus capitis. Its origin is the mid-dorsal line of the neck and fascia. The insertion is the superior nuchal line and atlas. It raises or turns the head.

Serratus Ventralis

The Serratus Ventralis is exposed by cutting the wing-like Latissimus Dorsi. The said muscle is covered entirely by adipose tissue. The origin is from the first nine or ten ribs and from part of the cervical vertebrae.

The insertion is the vertebral border of the scapula. It draws scapula forward, backward, and against the body.

Serratus Dorsalis

The Serratus Dorsalis is medial to both the scapula and the Serratus Ventralis. Its origin is via apoeurosis following the length of the mid-dorsal line, and its insertion is the dorsal portion of the last ribs. Its action is to draw ribs cranial.

Intercostals

The Intercostals are a set of muscles sandwiched among the ribs. They interconnect ribs, and are therefore the primary respiratory skeletal muscles. They are divided into the *external* and the *internal subscapularis*. The origin and insertion are in the ribs. The intercostals pull the ribs backwards or forwards.

Caudofemoralis

The Caudofemoralis is a muscle found in the pelvic limb and is unique to the felids (cats). The Caudofemoralis acts to flex the tail laterally to its respective side when the pelvic limb is bearing weight. When the pelvic limb is lifted off the ground, contraction of the Caudofemoralis causes the limb to abduct and the shank to extend by extending the hip joint.

Pectoral

Pectoantebrachialis: Pectoantebrachialis muscle is just one-half inch wide, and is the most superficial in the pectoral muscles. Its origin is the manubrium of the sternum, and its insertion is in a flat tendon on the fascia of the proximal end of the ulna. Its action is to draw the arm towards the chest.

Pectoralis Major

The pectoralis major, also called *pectoralis superficialis,* is a broad triangular portion of the pectoralis muscle which is immediately below the pectoantebrachialis. It is smaller than the pectoralis minor muscle. Its origin is the sternum and median ventral raphe, and its insertion is at the humerus. Its action is to draw the arm towards the chest.

Pectoralis Minor

The pectoralis minor muscle is larger than the Pectoralis major. However, most of its anterior border is covered by the pectoralis major. Its origins are ribs three–five, and its insertion is the coracoid process of the scapula. Its actions are the tipping of the scapula and the elevation of ribs three–five.

Xiphihumeralis

The most posterior, flat, thin, and long strip of pectoral muscle is the Xiphihumeralis. It is a band of parallel fibers that is found felines but not in humans. Its origin is the Xiphoid Process of the sternum. The insertion is the humerus.

Trapezius

In the cat there are three thin flat muscles that cover the back, and to a lesser extent, the neck. They pull the scapula toward the mid-dorsal line, anteriorly, and posteriorly.

Clavotrapezius

The most anterior of the trapezius muscles, it is also the largest. Its fibers run obliquely to the ventral surface. Its origin is the superior nuchal line and median dorsal line and its insertion is the clavicle. Its action is to draw the clavicle dorsally and towards the head.

Acromiotrapezius

Acromiotrapezius is the middle trapezius muscle. It covers the dorsal and lateral surfaces of the scapula. Its origin is the neural spines of the cervical vertebrae and its insertion is in the metacromion process and fascia of the clavotrapezius. Its action is to draw the scapula to the dorsal, and hold the two scapula together.

Spinotrapezius

Spinotrapezius, also called *thoracic trapezius*, is the most posterior of the three. It is triangular shaped. Posterior to the acromiotrapezius and overlaps latissimus dorsi on the front. Its origin is the neural spines of the thoracic vertebrae and its insertion is the scapular fascia. Its action is to draw the scapula to the dorsal and caudal regions.

Genitalia

Female Genitalia: In the female cat, the genitalia include two gonads, the uterus, the vagina, the genital passages and teats. Together with the vulva, the vagina of cat is involved in mating and provides a channel for newborns during *parturition*, or birth. The vagina is long and wide. Genital passages are the oviducts of the cat. They are short, narrow, and not very sinuous.

Male Genitalia

In the male cat, the genitalia includes the penis, which has a very similar surface texture to the tongue.

Cat Communication

Cat communication is the range of methods by which cats communicate with other cats, humans, and other animals. Communication methods include postures, movement (including "quick, fine" movements not generally perceived by human beings), and auditory and chemical signals.

Figure: *A primary method by which cats communicate is through body language, such as the position of the ears.*

The communication methods used by cats have been affected by the domestication process.

Auditory Communication Methods

Cats vocalize with purrs, growls, hissing and meows.

Meowing

A meow is a sound used by cats to signal a request to their mother or to a human being. Adult cats do not meow to each other, and so the meowing to human beings that domesticated cats exhibit is likely partly an extension of the use by kittens of this plaintive signal. The word *meow* is itself an onomatopoeia with various languages having different words for the sound including miau (Finnish, Lithuanian, German, Polish, Russian, Portuguese, Romanian and Spanish), miaou (French), nya (Ë0ã0, Japanese), miao (µU, Chinese - Mandarin, Italian), miau (Malay), ya-ong |Å9Æ (Korean), and meo-meo (Vietnamese).

Purring

A purr is a sound made by most species of felines. A tonal buzzing can characterise differently between cats. Domestic cats purr in a frequency of 25 to 150 vibrations per second. Purring is often understood as signifying happiness; however, cats sometimes purr when they are ill, or during tense, traumatic, or painful moments.

There is no unique anatomical feature that is clearly responsible for the sound.

Other Noises

Figure: *A cat hissing and arching its back to make itself appear larger to ward off a threat.*

Most cats growl or hiss when angered or feeling threatened, which serves as a warning to the offending party. If the warning is not heeded, a more serious attack may follow. Some may engage in behaviour or batting with their paws, with claws either extended or retracted. Cats sometimes make chirping or chattering noises when observing prey.

The cry of a cat in heat is called a caterwaul.

Body Language

Cats communicate a variety of messages using body language. Examples include arching their backs as a signal of fear or aggression, and slowly blinking to signal relaxation. A cat that chooses to lie with its stomach and chest exposed conveys trust, and comfort (this is also typical of overweight cats, as it is more comfortable for them); however, a cat may also roll on its side or back to be able to defend itself with all four sets of claws. Usually other signs (like ears and whiskers folded backwards) give an indication of the cat's overall mood. Flattened ears mean that the cat feels threatened, and may attack. Mouth open and no teeth exposed suggests a feeling of playfulness.

Figure: *A surprised cat*

The tail is often used as a signaling mechanism. A tail held high suggests happiness or confidence, or it can be used as a friendly greeting towards human beings or other cats (usually close relatives), while a half-raised tail shows less pleasure, and unhappiness is indicated with a tail held low. In addition, a cat's tail may "wag" or move rapidly to express a state of conflict. A cat with its tail held high and twitching shows excitement, but this is often mistaken for anger. Cats will twitch the tips of their tails when hunting or when irritated, while larger twitching indicates displeasure. They may also twitch their tails when playing. A scared or surprised cat may puff up its tail, and the hair along its back may stand straight up and the cat will turn its body sideways to a threat, in order to increase its apparent size. Tailless cats, such as the Manx, which possess only a small stub of a tail, move the stub around as if they possess a full tail.

Touching noses is a friendly greeting for cats, while a lowered head is a sign of submission. Some cats will rub their faces along their guardian's cheek, hands, or ankles as a friendly greeting or sign of affection. This action is also sometimes a way of "marking their territory", leaving a scent from the scent glands located in the cat's cheeks. More commonly, a cat will do a "head bonk" (or "bunt"), i.e., bump someone with the front part of its head to express affection.

Cats also lick each other and people (e.g. their owners). Cats lick each other to groom one other and to bond (this grooming is usually done between cats who know each other very well). They will also sometimes lick people for similar reasons. These reasons include wanting to "groom" people and to show them care and affection.

Kittens tend to follow one another, even into traps. This is because they don't feel confident enough to venture out on their own, without a companion. "You see, that's what makes kittens so easy to herd," said Chief Cat-Herder Robert Almanza.

Cats may paw their human companions, or a soft object on which they may be sitting, with a kneading motion. Cats often use this action alongside purring to show contentment and affection for their companions. This can also indicate curiosity. A cat may also do this when in pain or dying, as a method of comforting itself. It is instinctive to cats, and they use it when they are young to stimulate the mother cat's breast to release milk during nursing. As a result, cats hand-raised by human beings may not exhibit this behaviour. Pawing is also a way for cats to mark their territory. The scent glands on the underside of their paws release small amounts of scent onto the person or object being pawed, marking it as "theirs" the same way they would urinate to mark their territory. Since the nature of the activity is an instinctive response related to the mother's care for the kitten, it may be an expression of need, indicating an empty water bowl, hunger, an unappealing litter box, or the need for some attention from the caregiver.

Biting

Although a gentle bite can signify playfulness, bites that are accompanied by hissing or growling do not signify playful behaviour. During play, a cat can become overexcited, which can result in bites that, while intended only to be playful, can be strong enough to draw blood.

When cats mate, the male tom bites the scruff of the female's neck as she assumes a position conducive to mating.

Scent

Cats can communicate through scent via urine, feces, and chemicals in skin glands located around the mouth, tail, and paws. They also use scent in order to mark their territory. If another animal tries to get in the cat's territory, it will fight for the territory, or the cat will scare the animal off. Urine spraying is also a territorial marking. Cats rub up against furniture or doorways to mark the items as "theirs". When cats rub people, they are marking them with their scent, claiming them as "theirs". In addition, they are picking up the peoples' scents.

Cat Behaviour

Cat behaviour generally refers to the behaviours and habits of domestic cats, including body language and communication. Cat behaviour may vary among breeds and individual cats.

Many common behaviours include hunting techniques and reactions to certain events as well as interactions with humans and other animals, such as dogs. Communication can vary greatly depending on a cat's temperament. In a family with multiple cats, social position can also affect behaviour patterns with others.

A cat's eating patterns can also vary depending on the owner's choice of food or eating times/quantities. In the case of a family having two or more cats, one cat may become dominant over the other cats.

Body Language

Cats rely strongly on body language to communicate. Rubbing against an object, licking, and purring (sometimes) show affection. Purring can also mean that they are hurt or in pain.

Scent Rubbing and Spraying

This behaviour is used primarily to claim ownership of something, although unlike male cats, female cats do not usually spray. Some spayed females may spray to mimic males, but do not have enough scent to mark. Once male cats are neutered, the *scent rubbing* or *spraying* will, in most cases, decrease or stop. Some male cats continue to spray if not neutered early enough, though sometimes they continue to spray even if they were neutered early in life.

Kneading

Figure: *A cat kneading a soft blanket*

Kneading is an activity common to all domestic cats whereby, when in a state of ease, they alternately push out and pull in their front paws, often alternating between right and left limbs. Some cats will actually appear to "nurse" or suck on clothing during kneading.

History

Multiple theories exist that explain why cats knead. Kneading may have an origin going back to cats' wild ancestors who had to tread down grass or foliage to make a temporary nest in which to rest. Alternatively, the behaviour may be a remnant of a newborn's kneading of the mother's teat to stimulate milk excretion. Since most of the preferred "domestic traits" are neotenous or juvenile traits that persist in the adult, kneading may be a relict juvenile behaviour that is not lost in modern adult domestic cats.

Kneading is often a precursor to sleeping.

Many cats purr while kneading. They also purr mostly when newborn, when feeding, or when trying to feed on their mother's teat. The common association between the two behaviours may corroborate the evidence in favourof the origin of kneading as a remnant instinct. Some experts consider kneading to "stimulate" the cat and make it feel good, in the same manner as a human stretching.

Action

The cat exerts firm downwards pressure with its paw, opening its toes to expose its claws, then closes its claws as it lifts its paw. The process takes place with alternate paws at intervals of one to two seconds. They may do this while sitting on their owner's lap, which may prove painful if the cat is large or strong or has sharp claws (as the claws tend to dig into one's lap). Though cats will sit happily on a hard surface, they will only knead a soft or pliant surface, although some cats will reflexively "march" on hard surfaces instead of kneading them.

Courting

Cats, compared to many other mammals, have a unique courting style. The first step in courtship is the female coming into season, or "heat". Male cats will be able to smell a female cat in heat miles away and will therefore seek her out. Males will fight mercilessly for the right to be the first to mate with the female. After the dominant male has left, the less dominant males will then each mate with the female in turn. It is therefore possible that even if a male cat loses first breeding rights, he can still be the father. This is also the reason that a litter of kittens can have two or sometimes even three fathers.

Vocal Calls

- Purring – Purring is often a sign of contentment. Some cats purr when they are in extreme pain, or in labour, simply to try to calm themselves down. Purring therefore can be a sign of pleasure or pain; usually it is the former. Scientists have not yet been able to discover how purring works, but it is suspected that it is caused by minute vibrations in the voice box.
- Greeting – A particular sort of vocalisation, such as a low meow or chirp, possibly with simultaneous purring.
- Distress – Mewing is a plea for help or attention often made by kittens. There are two basic types of this call, one more loud and frantic, the other more high-pitched. In older cats, it is more of a panicky repeated meow.
- Attention – Often simple meows and mews in both older cats and young kittens. A commanding meow is a command for attention, food, or to be let out.
- Protest – Whining meows.
- Frustration – A strong sigh or exhaled snort.
- Happy – A meow that starts low then goes up and comes back down.
- Watching/interest – Cats will often "chatter" or "chirrup" on seeing something of interest. This is sometimes attributed to mimicking birdsong to attract prey or draw others' attention to it, but often birds are not present.

Panting

Unlike dogs, panting is an uncommon occurrence in cats. However, some cats can pant when stressed, such as a car ride. Most commonly, cats will pant in response to environmental changes, such as anxiety, fear, excitement, or heat. However, if panting is excessive or the cat appears in distress, it becomes important to identify the underlying cause, as panting may be a symptom of a more serious condition, such as a nasal blockage, heartworm disease, head trauma, or drug poisoning. Other problems, such as fatigue, weight loss, poor appetite, excessive drinking, vomiting, or diarrhea, may also be present. If the panting appears to be in response to heat or fear, the owner should remove the stimulation and continue to observe his or her pet. If panting continues, the owner should consult a veterinarian. In many cases, feline panting, especially if accompanied by other symptoms, such as coughing or shallow breathing (dyspnea), is considered to be abnormal, and treated as a medical emergency.

Food Eating Patterns

Cats are obligate carnivores, and can survive without vegetation. Felines in the wild will usually hunt smaller mammals regularly throughout the day to keep themselves nourished. Domestic cats, however, are used to a relaxed lifestyle and, therefore, will eat even smaller amounts, but more regularly. Many cats will find and chew small quantities of long grass but this is not for its nutritional value, it is a purely mechanical function. The eating of grass triggers a regurgitation reflex to help expel indigestible matter, like hairballs and the bones of prey.

Socialisation

Kittens are naturally afraid of people at first, but if handled and well-cared for in the first 16 weeks, they will develop trust in the humans who care for them. To decrease the odds of a cat being unsocial or hostile towards humans, kittens should be socialised at an early age. It is a challenge to socialise an adult feral cat. Socialised adult feral cats tend to trust only those people they have learned over time can be trusted, and can be very fearful around strangers.

Cats can be extremely friendly companions. The strength of the cat–human bond is mainly correlated with how much consideration is given to the cat's feelings by his human companion. The formula for a successful relationship thus has much in common with human to human relationships.

Some people regard cats as sneaky, shy, or aloof animals. Cats have an inherent distrust for predator species such as humans, and often seek to minimise any contact with people they do not perceive as trustworthy. Feline shyness and aggression around people with cat social skills is often a result of lack of socialisation, abuse or neglect. Cats relate to humans differently than more social animals, enjoying some time on their own each day as well as time with humans.

Cats have a strong "escape" instinct. Attempts to corner, capture or herd a cat can thus provoke powerful fear-based escape behaviour. Socialisation greatly reduces the number of humans that a cat will respond to in this way. Socialisation is a process of learning that many humans can be trusted. There is a widespread belief that relationships between dogs and cats are problematic. However, both species can develop amicable relationships by reading each others' body language correctly. The animals can better read each others' language when they first encounter each other at a young age, due to the fact that they are learning to communicate simultaneously. The order of adoption may also cause significant differences in their relationship. Sometimes, the dog may be simply looking to play with the cat while the cat may take the approach a lot more seriously and lash out with its sharp claws, causing painful injury. Such an incident may cause an irreversible animosity between the cat and dog.

Cat Senses

Cat senses are adaptations that allow cats to be highly efficient predators. Cats have acute sight, hearing and smell, and their sense of touch is enhanced by long whiskers that protrude from their heads and bodies. These senses allow cats to hunt effectively in dim light or at night.

Sight

Testing indicates that a cat's vision is greater at night in comparison to humans, and inferior in daylight. Cats, like dogs and many other animals, have a *tapetum lucidum*, which is a reflective layer behind the retina that sends light that passes through the retina back into the eye. While this improves the ability to see in darkness, it appears to reduce net visual acuity, thus detracting when light is abundant. In very bright light, the slit-like iris closes very narrowly over the eye, reducing the amount of light on the sensitive retina, and improving depth of field. Big cats have pupils that contract to a round point. The tapetum and other mechanisms give the cat a minimum light detection threshold up to seven times lower than that of humans. Variation in colour of cats' eyes in flash photographs is largely due to the reflection of the flash by the tapetum.

***Figure:** The tapetum lucidum reflecting green in the pupils of a cat.*

Many cats have a visual field of view estimated at 200°, versus 180° in humans, with a binocular field (overlap in the images from each eye) narrower than that of humans. As with most predators, their eyes face forward, affording depth perception at the expense of field of view. Field of view is largely dependent upon the placement of the eyes, but may also be related to the eye's construction. Instead of the fovea, which gives humans sharp central vision, cats have a central band known as the visual streak. Cats can see some colours, and can tell the difference between red, blue and yellow lights, as well as between red and green lights. Cats are able to distinguish between blues and violets better than between colours near the red end of the spectrum.

Cats have a third eyelid, the nictitating membrane, which is a thin cover that closes from the side and appears when the cat's eyelid opens. This membrane partially closes if the cat is sick, although in a sleepy, content cat this membrane is often visible. If a cat chronically shows the third eyelid, it should be taken to a veterinarian for evaluation.

Unlike humans, cats do not need to blink their eyes on a regular basis to keep their eyes lubricated (with tears). Unblinking eyes are probably an advantage when hunting. Cats will, however, "squint" their eyes, usually as a form of communication. Cat owners can often entice their pets to squint or even fully close their eyes just by talking to them in a soothing or pleasing manner. Many cats will also squint in response to seeing their owners squint. When a cat does blink, it is a slower blink than in humans, so it will typically blink one eye at a time so it can still see from the other eye. When the cat is looking at someone while doing this, it gives the impression that the cat is winking at them.

Cats have a wide variation in eye colour, the most typical colours being golden, green and orange. Blue eyes are usually associated with the Siamese breed, but they are also found in white cats and kittens. It is a common misconception that all white cats with blue eyes are deaf. This is not true, as there are many blue-eyed cats with perfect hearing. However, white cats with blue eyes do have slightly higher incidences of genetic deafness than white cats of other eye colours. White cats having one blue and one other-coloured eye are called "odd-eyed" and may be deaf on the same side as the blue eye. This is the result of the yellow iris pigmentation rising to the surface of only one eye, as blue eyes are normal at birth before the adult pigmentation has had a chance to express itself in the eye(s).

Hearing

Humans and cats have a similar range of hearing on the low end of the scale, but cats can hear much higher-pitched sounds, up to 64 kHz, which is 1.6 octaves above the range of a human, and even 1 octave above the range of a dog. When listening for something, a cat's ears will swivel in that direction; a cat's ear flaps (pinnae) can independently point backwards as well as forwards and sideways to pinpoint the source of the sound. Cats can judge within three inches (7.5 cm) the location of a sound being made one yard (approximately 91 cm) away—this can be useful for locating their prey.

Smell

A domestic cat's sense of smell is about fourteen times as strong as a human. Cats have twice as many receptors in the olfactory epithelium (i.e. smell-sensitive cells in their noses) as people do, meaning that cats have a more acute sense of smell than humans. Cats also have a scent organ in the roof of their mouths called the vomeronasal (or Jacobson's) organ. When a cat wrinkles its muzzle, lowers its chin, and lets its tongue hang a bit, it is opening the passage to the vomeronasal. This is called *gaping*, "sneering", "snake mouth", or "flehming". Gaping is the equivalent of the Flehmen response in other animals, such as dogs, horses and big cats.

Touch

A cat has about twenty-four movable vibrissae ("whiskers"), in four rows on each upper lip on either side of its nose (some cats may have more). There are also a few on each cheek, tufts over the eyes, bristles on the chin, the cat's inner "wrists", and at the back of the legs. The Sphynx (a nearly hairless breed) may have full length, short, or no whiskers at all.

The structure of the brain region (barrel cortex) which receives information from the vibrissae is similar to that found in the visual cortex which permits the cat to create a three dimensional map of its surroundings. This doesn't mean that sensing with vibrissae is a type of vision. It is still a touch sensation and environmental information is built up incrementally (in small steps).

Vibrissae aid sensation and navigation. The upper two rows of whiskers are able to be moved independently from the lower two rows for greater precision during measurement. A cat's whiskers are more than twice as thick as ordinary cat hairs, and their roots are three times deeper in a cat's tissue than other hairs. They have numerous nerve endings at their base, which give cats extraordinarily detailed information about nearby air movements and objects with which they make physical contact. They enable a cat to know that it is near obstacles without it needing to see them.

Whiskers also aid in hunting. High speed photography reveals that when a cat is unable to see its prey because it is too close to its mouth, its whiskers move so as to form a basket shape around its muzzle in order to precisely detect the prey's location. A cat whose whiskers have been damaged may bite the wrong part of its prey indicating that they provide cats with detailed information about the shape and activity of its prey. Whiskers may also help cats to detect scents more readily by directing air currents to their nose and mouth.

Whiskers are also an indication of the cat's attitude. Whiskers point forward when a cat is inquisitive and friendly, and lie flat on the face when the cat is being defensive or aggressive.

Taste

The cat family was shown in 2005 to lack the T1R2 protein, one of two required for function of the sweetness sensory receptor; a deletion in the relevant gene (*Tas1r2*) causes a shift in the genetic reading frame, leading to transcription stopping early and no detectable mRNA or protein produced. The other protein, T1R3, is present and identical to that of other animals, and the relevant taste buds are still present but inactive. Such a genetic marker found in the entire family and not other animals must be the result of a mutation in an early ancestor; as a deletion mutation it could not revert, and thus would be inherited by all descendants, as the evolutionary tree branched out.

Some scientists now believe this is the root of the cat family's extremely specialised evolutionary niche as a hunter and carnivore. Their modified sense of taste would cause them to some degree to ignore plants, a large part of whose taste appeal derives from their high sugar content, in favourof a high-protein carnivorous diet, which would still stimulate their remaining taste receptors.

Cat Righting Reflex

The cat righting reflex is a cat's innate ability to orient itself as it falls in order to land on its feet. The righting reflex begins to appear at 3-4 weeks of age, and is perfected at 7 weeks. They are able to do this as they have an unusually flexible backbone and no functional clavicle (collarbone). The minimum height required for this to occur in most cats (safely) would be around 1 m (3.3 ft). Cats without a tail also have this ability, since a cat mostly moves its hind legs and relies on conservation of angular momentum to set up for landing, and the tail is in fact little used for this feat.

Technique

After determining up from down visually or with their vestibular apparatus (in the inner ear), cats manage to twist themselves to face downward without ever changing their net angular momentum. They are able to accomplish this with these key steps:

1. Bend in the middle so that the front half of their body rotates about a different axis than the rear half.
2. Tuck their front legs in to reduce the moment of inertia of the front half of their body and extend their rear legs to increase the moment of inertia of the rear half of their body so that they can rotate their front half quite far (as much as 90°) while the rear half rotates in the opposite direction quite a bit less (as little as 10°).
3. Extend their front legs and tuck their rear legs so that they can rotate their rear half quite far while their front half rotates in the opposite direction quite a bit less.

Depending on the cat's flexibility and initial angular momentum, if any, the cat may need to repeat steps two and three one or more times in order to complete a full 180° rotation.

Terminal Velocity

In addition to the righting reflex cats have a number of other features that will reduce damage from a fall. Their small size, light bone structure, and thick fur decrease their terminal velocity. Furthermore, once righted they may also spread out their body to increase drag and slow the fall to some extent. A falling cat's terminal velocity is 100 km/h (60mph) whereas that of a falling man in a "free fall position" is 210 km/h (130mph). At terminal velocity they also relax as they fall which protects them to some extent on impact. However, it has been argued that, after having reached terminal velocity, cats would orient their limbs horizontally such that their body hits the ground first.

Injury

Using their righting reflex theory, cats can often land uninjured. This is, however, not always the case, and cats can still break bones or die from falls. In a 1987 study, published in the *Journal of the American Veterinary Medical Association*, of 132 cats that were brought into the New York Animal Medical Center after having fallen from buildings, it was found that the injuries per cat increased depending on the height fallen up to seven stories but decreased above seven stories. The study authors speculated that after falling five stories the cats reached terminal velocity and thereafter relaxed and spread their bodies to increase drag.

However, an alternative interpretation which came out of internet chat of the study would be that upon an excess of seven stories the cats experience a higher fatality rate which precludes the owner from bringing them in for medical attention. Although scientists in Massachusetts have recently discovered that the cat's ability to spread its legs out to decrease drag when reaching terminal velocity would explain the decreased injuries sustained above seven stories because they wouldn't reach terminal velocity before then.

Professor David Stevenson said "we simulated the cat's weight and size and found the terminal velocity to be 60mph which would more than likely result in severe injury or death to the cat when falling from this speed, but once we took into account the cat's ability to right its self and spread its body out this reduced the terminal velocity to only 53mph. this 7mph difference is massive and would almost certainly ensure the cat's survival. There however is always the possibility that the cat may not manage to right itself so this is far from a conclusive experiment and we do not condone the throwing of cats from anything".

Falling Cat Problem

The falling cat problem consists of explaining the underlying physics behind the common observation of the cat righting reflex: how a free-falling cat can turn itself right-side-up as it falls, no matter which way up it was initially, without violating the law of conservation of angular momentum.

Although somewhat amusing, and trivial to pose, the solution of the problem is not as straightforward as its statement would suggest. The apparent contradiction with the law of conservation of angular momentum is resolved because the cat is not a rigid body, but instead is permitted to change its shape during the fall. The behaviour of the cat is thus typical of the mechanics of deformable bodies.

The solution of the problem, originally due to (Kane & Scher 1969), models the cat as a pair of cylinders (the front and back halves of the cat) capable of changing their relative orientations. Montgomery (1993) later described the Kane–Scher model in terms of a connection in the configuration space that encapsulates the relative motions of the two parts of the cat permitted by the physics. Framed in this way, the dynamics of the falling cat problem is a prototypical example of a nonholonomic system (Batterman 2003), the study of which is among the central preoccupations of control theory. A solution of the falling cat problem is a curve in the configuration space that is *horizontal* with respect to the connection (that is, it is admissible by the physics) with prescribed initial and final configurations. Finding an optimal solution is an example of optimal motion planning (Arbyan & Tsai 1998; Xin-sheng & Li-qun 2007).

In the language of physics, Mongomery's connection is a certain Yang-Mills field on the configuration space, and is a special case of a more general approach to the dynamics of deformable bodies as represented by gauge fields (Montgomery 1993; Batterman 2003), following the work of Shapere and Wilczek (Shapere and Wilczek

Buttered cat Paradox

The buttered cat paradox is a paradox based on the tongue-in-cheek combination of two adages:

- Cats always land on their feet.
- Buttered toast always lands buttered side down.

The paradox arises when one considers what would happen if one attached a piece of buttered toast (butter side up) to the back of a cat, then dropped the cat from a large height. The buttered cat paradox, submitted by artist John Frazee of Kingston, New York, won a 1993 *OMNI* magazine competition about paradoxes.

Thought Experiments

Some people jokingly maintain that the experiment will produce an anti-gravity effect. They propose that as the cat falls towards the ground, it will slow down and start to rotate, eventually reaching a steady state of hovering a short distance from the ground while rotating at high speed as both the buttered side of the toast and the cat's feet attempt to land on the ground. In June 2003, Kimberly Miner won a Student Academy Award for her film *Perpetual Motion.* Miner based her film on a paper written by a high-school friend that explored the potential implications of the cat and buttered toast idea.

In Humor

The faux paradox has captured the imagination of science-oriented humorists. Testing the theory is the main theme in an episode of the comic book strip *Jack B. Quick*, the title character seeks to test this theory, leading to the cat hovering above the ground, with the cat's wagging tail providing propulsion. The March 31, 2005, strip of the webcomic *Bunny* also explored the idea in the guise of a plan for a "Perpetual Motion MoggieToast 5k Power Generator", based on Sod's Law. In *Science Askew*, Donald E. Simanek comments on this phenomenon.

The idea appeared on the British panel game *QI*, where the idea was discussed. As well as talking about the idea, they also brought up other questions regarding the paradox. These included "Would it still work if you used margarine?", "Would it still work if you used I Can't Believe It's Not Butter?", and "What if the toast was covered in something that was not butter, but the cat thought it was butter?", the idea being that it would act like a placebo.

Chapter 7

Rat Behaviour

The Actors

The rats who have performed most of these behaviours for me are my three male rats, Cricket, Widget and Snip. Here they are at 7 months old:

To tell them apart in the following photographs:

- Snip is the black hooded rat (white with a black head and broken dorsal stripe).
- Cricket is a black berkshire (black with a white belly), with a white stripe on his forehead and smooth fur.
- Widget is a black berkshire too, but he has wavy rex fur. His head markings are also different from Cricket's, for Widget has

an all-black forehead and a white spot on the left side of his nose. Widget also has a mild head tilt, which does not bother him but causes his head to tilt slightly to his right.

Social Behaviour

Agonistic Behaviour

A quick note: "agonistic behaviour" refers to the complex of aggression, appeasement and avoidance behaviour that occurs between members of the same species. Agonistic behaviour is a much broader term than "aggression," which refers to behaviour patterns which serve to intimidate or damage another (for more, see McFarland, 1982).

For a more detailed article on this topic, see Aggression in Norway rats.

Social Agonistic Behaviour

Rats are social animals, and they establish and maintain simple dominance hierarchies through social agonistic behaviour. Such hierarchies generally consist of one dominant individual and several subordinates. Most social agonistic behaviour in pet rats is seen between males, but it may also occur between females. In males, social aggression increases around age 6 months.

Agonistic Behaviour Sequences

Agonistic behaviours are strung together into sequences that vary in their intensity and duration. The lowest intensity encouters are chases. As the intensity increases, you may see stand-offs and physical contact like boxing and sidling. These physical encounters may escalate in rare instances into a fight.

The immediate "goal" of much offensive behaviour is for one rat to deliver a bite to the subordinate rat's rump, while the "goal" of the defending rat is to prevent such a bite.

Aggressive neck grooming is a form of agonistic behaviour. Grooming consists of rapid little nibbles in which the groomer seizes folds of neck skin between his teeth (Miczek and Boer 2004). The groomed rat remains immobile and may peep or squeak softly. Any sudden movement by the groomed rat may trigger a bite and kick from the groomer.

Many agonistic interactions consist of one rat chasing another for a few seconds. The chased rat flees and either outruns or hides from the pursuer, or sometimes the pursuer desists. If the pursuer catches up, he may nip or bite the rump of the fleeing rat and may attempt to engage him in an encounter. Sometimes, the pursuer may mount the fleeing rat. The aggressing rat may also engage in aggressive neck grooming.

If the chased rat holds his ground, chaser and chasee may have an encounter. A pursued rat may turn to face the pursuer, initiating a nose-off. The defending rat may show an open-mouth tooth display, with long squeaks and sometimes hisses. Sometimes the tails of both rats writhe on the ground, and both rats' fur may be piloerect. Frequently the encounter goes no further, and one of the rats (usually the pursued one) flees.

If neither rat flees, the encounter may escalate with physical contact. The rats may box and/or sidle, and if the sidling rat gets very close he may push or kick out with a hind foot. The subordinate rat, especially if he is young or the rats are confined in a cage, may roll over into a belly-up roll. I've found among my rats that most of the few encounters that get this far end here, with the flight of the pursued rat. After the flight another sequence may occur, and on a few occasions combinations of chases, sidling, and flight have lasted for tens of minutes or even an hour or more (these tense times are known as "Widget rampages").

A variation on the sidle sequence is seen on ledges such as shelves and hammocks. On ledges, the sidling rat crowds the other toward the edge. This may be a very slow process, with the sidling rat maneuvering a fraction of an inch at a time, with long pauses when nobody moves. If the second rat does not jump off the ledge himself, he will be pushed off. Sometimes the pursuer will follow the second rat around the cage, pushing him off ledge after ledge. If the second rat resists, the encounter may escalate.

Rarely, physical contact may escalate into a fight. On a handful of occasions the encounter escalates beyond the sidle-and-kick stage, and the two rats close into a fighting ball, in which the attacking rat may bite the other's flank or rump. These fights are usually brief (lasting just a few seconds) and end with one rat fleeing, hiding and staying staying quiet and subdued for a while — an hour or more.

For more on rat aggression, see Blanchard *et al.* 1975, Blanchard and Blanchard 1977, Blanchard *et al.* 1977.

Observations of Agonistic Behaviour

Their first year: Widget emerged as the dominant rat in the cage. Widget directed most of his aggression toward Snip and chases were common. During a chase Snip typically retreated to the cage, which has a single door in the roof. Widget usually stopped his pursuit before climbing the cage wall (with his head tilt, Widget is not as agile a climber as Snip). If Widget chased him out of the cage, Snip had favorite hiding places in the room. His first hiding place was on top of a computer that rested on the floor. One day, Widget managed to climb the computer and forcibly pushed Snip off the computer. Both rats fell, fighting, to the floor. They separated on impact and neither was hurt, but Snip never retreated to the computer again, his safe spot had been violated. He chose my office chair as a new hiding place and and rested high up there where Widget couldn't reach him.

However, sometimes Snip actually seemed to seek Widget out to trigger an encounter. On at least two occasions, Widget's "rampage" was dying down, and Snip appeared to prolong the conflict. Snip blocked Widget's exit from the nestbox, confronted him elsewhere in the cage, and then followed Widget closely. Snip did not employ dominant tactics, however. No sidling or lateral pushing or biting Widget's rump. Intstead, Snip's actions triggered close nose-offs and boxing encounters, during which Snip repeatedly nipped Widget's nose. Sometimes Snip pushed Widget over with his front feet while boxing, and groomed and nipped Widget's belly, while Widget squeaked short squeaks. After fifteen minutes of this, Widget went seriously on the offensive again, chased Snip around the cage and cornered him. Snip emitted long squeaks and 20 kHz ultrasound squeaks (barely audible) and hissed and both rats chattered their teeth, then Widget attacked and bit Snip. Snip's use of defensive tactics in these encounters with Widget is not because Snip does not know offensive tactics. I have seen Snip use offensive tactics against Cricket several times. Snip chases, sidles, and kicks at Cricket... he just never does this to Widget.

During this first year, Widget bit Snip twice: at age 4.5 months he bit Snip on the rump, and at age 7 months he bit Snip on the side of the belly. Both wounds healed uneventfully. I noticed that these bites tended to occur when Snip was cornered and could not escape Widget (e.g. inside a nestbox, corner of a high shelf). I responded by eliminating places where a rat could get cornered: I cut four holes in each nestbox so a rat inside could escape from any angle, and I slung a hammock below the high shelf so a cornered rat could always jump down to safety. Widget continued to chase Snip, but Snip was rarely cornered again.

Age one year and beyond: As the rats approached one year of age, these aggressive encounters escalated, especially between Widget and Snip. I heard distressed squeaking and saw boxing and sidling almost every day. During free-range time, Widget chased Snip and Cricket so much that they spent all their time hiding in the cage while Widget had the run of the entire floor.

At 11.5 months, things came to a head. Widget bit Cricket on the scrotum, opening an inch long gash. While Cricket was recovering in another cage, Widget bit Snip twice on the hind legs (on the knee and lower leg), drawing blood, and in self defenceSnip bit Widget on the face just under the eye. After these bites I separated Widget permanently from the other two rats. He now lives in a separate cage next door to theirs. Snip and Cricket get along well, displaying only mild aggression, and have never bitten each other.

Widget now has free range time at a different time from the two other rats. Snip and Cricket, instead of hiding in the cage, now spend their free range time exploring and running about.

Interestingly, during free range times in the months after Widget's separation, Widget and Snip still engaged in nose-offs through the bars of the cage (Widget sought out Snip, Snip sought out Widget). Snip would squeak piteously, and frequently hissed. They sometimes lunged at each other as well, but could not make contact due to the cage bars (for pictures of these between-the-bars altercations, see the photos under hiss)

Snip, the omega rat in the original threesome (Cricket was a beta) became the dominant rat in the little Snip-Cricket twosome. Snip chases Cricket and sometimes sidles and kicks him, but these encounters are uncommon, brief, and not very intense. Cricket emits a few squeaks sometimes but rarely seems perturbed.

Snip and Cricket have fought only once since Widget's separation, when Widget had a nose-off with Snip through the cage bars. Snip became distressed, displaying piloerect fur and prolonged squeaking for several minutes. Snip then turned and attacked Cricket. Nobody was hurt, but after incident I put both rat cages on stands, raising them about a foot off the ground, making such between-the-bars encounters impossible. There have been no encounters and no fights since then.

Maternal Aggression

A different category of rat aggression is maternal aggression, in which a pregnant or lactating mother rat attacks intruders (other rats, humans, or other animals).

Maternal aggression can include a lunging attack, typically directed at the neck or back of an intruding rat. The mother rat may also bite, sidle, and kick (thrust of one hind leg toward intruder during a sidle). Intruders usually respond with nose-offs or belly-up postures (Erskine *et al.* 1978). Mother rats may also attempt to drag the intruder by the neck or side (Price and Belanger 1977).

Maternal aggression is present but occurs relatively infrequently at the end of pregnancy, and starts to increase after the birth of the litter. Maternal aggression peaks during the 9th day of lactation. On the ninth day, the highest percentage of females attacks and bites intruders, with the lowest hesitation before doing so. After the ninth day, maternal aggression declines (Erskine *et al.* 1978).

The presence of the litter is an important in triggering maternal aggression: if the litter is removed the mother rat's aggression subsequently decreases. Four hours after the litter's removal, the mother rat's aggression is down to nearly normal levels (Erskine *et al.* 1978).

Maternal aggression may discourage other animals from approaching and possibly harming her offpsring (e.g. predators, unfamiliar rats who could commit infanticide).

Play Fighting

To me, as a rat owner, young rat play involved a lot of chasing, fleeing, rolling over, jumping on each other and, as they grew a bit older, boxing and sidling. There didn't appear to be too much pattern to it to me as a casual observer, with the three rats playing with each other in all combinations. Over time, however, one rat (Widget) appeared to come out on top more often. As the rats grew older, they started fighting seriously, using some of the same behaviours, as well as boxing, sidling, nipping and biting.

I got interested in rat play after my rats had stopped play-fighting so much, and so I did a search in the library to see what I could find. There is a wealth of information out there on rat play! Here's what I found:

Young rats play intensely with each other: they chase and flee, roll over and box, jump on each other and bounce around the cage. In fact, they use many of the same behavioural sequences we see later in agonistic encounters.

Play fighting in rats is different from adult fighting, however, in that it targets a different area of the body and may involve different tactics. Play fighting involves attack and defenceof the *nape*, while serious fighting involves attack — with intent to bite — and defenceof the *rump*.

Juveniles and subordinate adults defend against nape attacks by (among other tactics) rolling over into the supine position (belly-up roll) which hides the nape and rump from the attacker. The attacker may stand over the supine rat and may nibble at the sides of the belly (belly grooming), attempting to reach the nape. The pinned rat may peep and squeak.

Adult dominance hierarchies can be predicted from this juvenile play fighting: oddly enough, the most frequent juvenile play-attacker becomes the *subordinate* rat after sexual maturity. This reversal is due to the least-frequent-attackers' emerging proficiency at defenceand successful counterattack, namely, switching from the belly-up roll to the forequarter pivot when attacked.

Play fighting serves a social maintenance function in both juveniles and adults. Subordinates direct more playful contacts at dominant rat than at each other. Directing such playful attacks on the dominant rat may lead the dominant rat to tolerate the subordinate's presence. As adults, the dominant rat tends to evade these encounters with adult defencetactics (forequarter pivot).

Subordinate rats, when playfully attacked by the dominant rat, roll over into the juvenile defenceposition. (the supine rat is usually a subordinate, but I have occasionally seen the dominant on his back before a subordinate too). The recumbent rat may make a few efforts to get up and may be pushed back again. These encounters usually end when the top rat desists and the supine rat rolls onto his feet. Usually the encounter goes no further and the pinned rat does not flee once he gets to his feet. On one occasion, however, I observed a belly-groom escalate into a fight, when both rats were confined in a small travel cage.

Not all subordinates are the same. Some initiate many playful contacts with the dominant, and in turn tend to be tolerated by him. Others avoid the dominant and are more frequently attacked by him when he encounters them. These avoiders are more likely to rise to the dominant position if the dominant rat is removed. For those who are interested, here's a more detailed description of play in rats. And, for more on animal play in general, see Burghardt 1999.

- Barnett 1975
- Blanchard and Blanchard 1977, 1990
- Panksepp, Jaak 1981
- Pellis and Pellis 1987, 1991, 1992
- Pellis, Pellis, and McKenna 1993.

Allogrooming

Allogrooming means grooming another individual, as opposed to autogrooming, or grooming oneself.

Rats groom each other in a number of contexts. Nibbling and snout-touching directed at the nape may be the consumatory phase of play fighting, and the nibbled rat may submit to the grooming or may protest with peeps and squeaks and may pull away. Sometimes, to protect his nape and rump, one rat may roll over, leaving his belly exposed to belly-grooming.

Head and body grooming, especially directed at the eyes and mouth, may serve a health function, as the groomed rat cannot reach these areas with his tongue and teeth. Head and body grooming may be an amicable social behaviour that maintains ties between members of a group (which is also a function of play fighting), and may help to spread a common scent. It may function as olfactory investigation of the sebum from the dorsal surface of the rat. Sometimes the groomed rat protests with peeps and squeaks.

In my own rats, I see most head and body grooming between my two subordinate rats, Snip and Cricket. Less frequently, I see head and body grooming between one of them and the dominant rat, Widget.

Communication

Vocal Communication

Audible rat sounds: Rats produce a variety of audible vocalisations, from soft peeps and squeaks to loud shrieks. Many of the vocalisations, such as peeps and short squeaks and their variants, occur during mild social interactions such as play fighting or head grooming. Most audible squeaks and peeps indicate protest. Long squeaks tend to indicate stronger protest, and hissing occurs during escalated conflict. Rats shriek during fights, when they are in pain, or to indicate very strong protest.

Rats also grind their incisors together in a soft, rhythmic fashion, called bruxing. The movement is caused by the the jaw muscles pulling the jaw rapidly and repeatedly upward so that the incisors grind against each other. During high intensity bruxing, the masseter muscle, which runs through the rat's eye socket, may vibrate the eyeball in and out in time with the jaw, a phenomenon known as "eye boggling." The muscular activity involved in bruxing is quite different from that seen in chewing, and may involve a separate neural pattern generator (Byrd 1997).

Rats may brux to sharpen their incisors, called *thegosis* (Murray and Sanson 1998). Rat incisors grow the rat's entire life, and are ground down through gnawing and bruxing. Anecdotally, rats brux during times of relaxation, while rat is alone or sometimes while being gently stroked. Rats also brux during stress and uncertainty (Rosales *et al.* 2002, Pohto 1979). Bruxing during stressful situations, such as tense social interactions, is often called chattering. Chattering may be louder than bruxing and may contain more sharp cracking sounds. However, it is unclear whether bruxing and chattering are distinct sounds, or whether they are the same sound given in different contexts. A spectral analysis would help answer this question.

To read more about bruxing and hear bruxing samples, go to the Norway Rat Vocalisations Page.

Ultrasonic vocalisations: Rats also emit a number of vocalisations in the ultrasonic range, above our level of hearing (>20 kHz). Ultrasonic vocalisations include infant distress calls (around 30-50 kHz), long distress calls (around 20 kHz), and short, positive high pitched chirps (around 50 kHz). These high-pitched chirps have generated a lot of interest: here's an article on "laughing in rats" as well as a summary article written by the researchers. We cannot hear these sounds, but they can be recorded and replayed at our level of hearing using a bat detector. For more, see Bclee's Rat Detector Project.

Twenty kHz is at the very upper limits of our range of hearing. During a tense nose-off between Widget and Snip, I bent down close to them and heard a series of barely audible, high-pitched squeaks which might have been long distress USVs. I was amazed — the rats were perfectly still and it looked like nothing was happening, but in reality a whole barrage of communication was going on that I could barely detect!

A month later, I was recording squeaks while Widget and Snip boxed and nosed-off in the cage. The encounter got more intense, Snip emitted long squeaks and hisses, and both rats chattered their teeth. Widget attacked and I immediately broke it up. Later, looking at sonograms of the recordings, I saw a whole series of long, high pitched squeaks just above 20 kHz, right at the top of what my microphone could detect. Snip had been squeaking in ultrasound.

Olfactory Communication

Sexually mature rats, especially males, deposit tiny droplets of urine on surfaces and objects they walk on, known as urine marking. Rats prefer to mark surfaces and objects they can readily walk on, and their marking behaviour increases dramatically when they encounter an unfamiliar odor (Hopp and Timberlake 1983). Male rats prefer to urine mark over the odors of other adult rats, and neutered animals mark very little (Brown 1975, 1977).

In general, urine marking is considered a form of advertisement of one's presence and a sex attractant. Doty (1974) hypothesizes that male rats advertise and the female selects her mate from among the advertisers. Female rats advertise themselves through urine marking as well, though at much lower rates than males. Females urine mark the most on the night before estrus (Birke, 1978; Calhoun 1962, p. 151). Male rats find the urine and sebaceous odors of females to be highly attractive (Sachs 1997; Pfaff *et al.* 1973) so the female may be advertising her availability and receptivity to potential mates.

Urine marking isn't the only kind of chemical communication between rats. Rats have scent glands on their flanks, and they rub this scent on objects in their environment (known as flank marking). Males flank mark more than females, and rats flank mark more when they can smell other rats (Peden and Timberlake 1990).

Reproductive Behaviour

Female Rats

Female rat ovarian cycle: The female rat's cycle is divided into several phases: proestrus, which lasts 12 hours, corresponds to the onset of mating behaviour, known as *behavioural estrus*, or *heat*. The female *ovulates* at the end of proestrus and thus enters estrus, which lasts 36 hours. After estrus she enters diestrus, a 48 hour period that is subdivided into *diestrus I* (first 24 hours) and *diestrus II* (second 24 hours). After diestrus, she enters proestrus again.

Proestrus	→	ovulation	→	estrus	→	diestrus
12 hours				36 hrs		48 hrs

Female rat mating behaviour: During proestrus, or behavioural estrus, the female solicits the male to prompt him into mounting her. She darts toward him and runs or hops away. She may repeat this approach-retreat sequence several times, sometimes wiggling her ears. She may also pause near him or run by him, and may intercept him in his pursuit of another female.

The male finds these solicitation behaviours very attractive, and follows the female. He mounts her, and the pressure he exerts on her flanks, lower back, and anogenital area triggers lordosis in the female rat (Nelson 1995).

Lordosis is the female mating posture, in which she stands stiff and immobile, her back arched downward, her rump elevated, her tail deflected to one side, and her vulva facing backwards. Without lordosis copulation is not possible. After copulation, the female leaves the mating posture. She may groom herself or the male, she may walk about, or rest (Nelson 1995).

Male Rats

Male hormones and reproductive behaviour: Male sex drive is expressed after puberty, when the testes become active and start to secrete a hormone called *testosterone*. Testosterone does not cause reproductive behaviour outright so much as it increases the liklihood of reproductive behaviour occuring in a particular context, such as in the presence of a receptive female. A male with high testosterone levels is more likely to copulate with a receptive female than a male with low testosterone levels. Castration, which removes the testes and thus the source of testosterone, greatly reduces sexual responsiveness.

Male rat mating behaviour: Prior to mounting, the male engages in *precopulatory behaviour*: he investigates the female, sniffs her mouth and anogenital region.

Then he *mounts* her, and if she is in heat, she will exhibit the lordosis posture, which makes intromission possible. *Intromission*, when the male reproductive organ enters that of the female, is accompanied by thrusting motions of the hindquarters. The male's spine may arch and his forepaws may lift off the female's back. Intromission is followed by *ejaculation*, the foreceful expulsion of seminal fluid from the male's body through the urethra. A *sperm plug* may be deposited in the vagina, which serves to block intromission by other males for a while, thus giving the current male's sperm more time to fertilize the female. After mating, the male may emit ultrasonic vocalisations, and he becomes sexually inactive and lethargic. He may groom himself, then lie down and sleep (Nelson 1995).

Parental Behaviour

Maternal Behaviour

Pregnant female rats choose a nest site and build a nest before the young are born. Once the young are born, mother rats must perform three parental behaviours if the young are to survive:

- lick their pups to clean them of amniotic fluid and to stimulate the elimination of wastes.

- adopt the nursing posture, crouching over the pups to facilitate nursing, and protect and warm them
- retrieve the pups if they leave the nest (e.g. Smith and Berkson 1973).

To facilitate the first nursing by newborn pups, the mother rat licks her vulva and and belly during birth, spreading amniotic fluid onto her belly. The odor of the amniotic fluid attracts the newborn rats to her nipples. As the pups age, the odor of their own saliva and that of their littermates and mother keeps attracting them back to the nipples.

A mother of young pups may also exhibit maternal aggression, characterised by attacks directed at any animal who approaches her nest and young. Such attacks may discourage other animals from approaching and possibly harming her offspring.

The onset of maternal behaviour is influenced by hormones and by experience with pups.

Hormones play an important role in influencing maternal behaviour, but the exact relationship between the many hormones involved and maternal behaviour is not yet fully understood. Prolactin, a hormone closely associated with lactation, is known to be important, but many other hormones are also involved (see Nelson 1995 for more).

Exposure to pups also plays a role. Maternal behaviour may be triggered in non-mother female rats through repeated daily exposure to foster pups (a process called *concaveation*). In females who have never had litters, this process takes 5-6 days. Females who have had litters in the previous weeks have much lower latencies, however, and behave maternally toward foster pups after a day (Bridges 1990).

Maternal behaviour is maintained by tactile stimulation. Many maternal behaviours involve the mouth (licking, pup retrieval, nest building), and are drastically reduced if the mother's muzzle is anesthetised (Stern 1989).

The nursing posture is maintained through the stimulation of the young rooting against the nipples, if the mother's nipples are anesthetised she will not adopt the nursing posture (Stern *et al.* 1992). If the pups' mouths are anesthetised so that they do not root up against the mother, then an unanesthetised mother will not adopt the nursing posture either (Stern and Johnson 1990).

As the pups grow older, there is more mutually initiated contact between them and their mother. Around 14-16 days of age, the pups begin to explore the world outside the nest. The pups return to the nest guided by a chemical signal produced by the mother from days 15-27 of lactation (Leon and Moltz 1972).

Over time, mother-offspring contact declines, and maternal care wanes. The waning of maternal care is in part regulated by the mother's body temperature and her need to dissipate heat. A mother crouching over a litter cannot dissipate heat. As the pups grow older, they retain heat more efficiently, so the mother spends shorter and shorter bouts of time in the nest with them and avoids overheating herself (Woodside and Leon 1980).

Wild vs. domestic maternal behaviour: Domestication has had relatively little impact on maternal behaviour: maternal behaviour of domestic rats is qualitiatively similar to that of wild rats. Both wild and domestic mothers nurse, retrieve pups, build nests, and display maternal aggression. There are some quantitative differences, however: wild rats are more efficient at pup retrieval (wild mothers retreive more pups after they are scattered, and retrieve them more quickly). Domestic mothers spend more time building their nest and build bigger nests in a short period of time. Wild rats end up building equally large nests, but it takes them longer.

Communal nesting: Two or more mother rats may also pool their litters in a single nest and raise them together. My own rats were the product of a communal nesting situation: three females and two males were relinquished to the local humane society and were accidentally combined in a single cage. All three females became pregnant and all five rats were taken in by a local rat rescue organisation. One of the females didn't lactate, so the remaining two mothers raised the three litters together. It was impossible to tell which litter was which, and impossible to know whether the parents were related to each other. My three male rats are therefore some combination of full siblings, half siblings, and unrelated.

Paternal Behaviour

Before Weaning: Schultz and Lore (1993) studied nesting in groups of three adult rats: two females and one male. The females gave birth to two litters sired by the male. The male did not kill or harm the baby rats, but he did not care for them either: the male never attended to them or provided parental care. In fact, he rarely approached them at all. If he did so, he was severely attacked by the mothers. Most of the males spent spent their time alone in a corner of the cage, and by the time the pups were weaned all the males had been wounded.

The presence of the father had no effect on the success of the litters: mothers with and without the fathers present gave birth to the same number of pups, and raised the same percentage of pups to weaning, and spent similar amounts of time with their pups.

After weaning: Almost all studies of parental behaviour in rats focus on the mother. Usually, the father isn't even present. The few studies that consider the father examine his behaviour during his pups' infancy, when he is marginalised by the mother. Almost no studies examine the interactions between adult males and their juvenile offspring after weaning, even though it is highly unlikely that juvenile rats are socialised in the absence of adult males in the wild.

Drews *et al.* (1982) examines social behaviour in family groups consisting of adult females, adult males, and their post-weaning young. The authors recorded the behaviour of all the rats in the family groups during three 8 days observation cycles starting when the rats were 25, 60, and 79 days old.

Rat families were quite amicable. There was no wounding or biting between any of the rats. Threat displays were extremely rare, and tended to be given by young males toward other young males. Juvenile behaviour graded seamlessly into adult behaviour without any qualitative change: the forms of the behaviours didn't change, only their frequencies.

On the whole, all the rats wrestled, boxed, pinned chased and mounted each other. However, the frequencies of these behaviours changed over time in the different age-sex groups.

Juveniles wrestled together a great deal. To a lesser extent, adults wrestled with juveniles. Only half the wrestling bouts had a discernable winner. Of these, adult males usually won. Juvenile males won next most often, followed by juvenile females. Adult females usually lost. Boxing was relatively rare, and winners were hard to discern, but once again when winners could be discerned adult males consistently won. Juvenile males won less over time, while juvenile females won more as they grew older. Chasing increased over time too. Young females chased more than adult females, but overall chasing was a male activity. By the end of the study adult males were doing most of the chasing. Mounting increased over time, and by the end of the study adult males were doing most of the mounting. Juveniles were more likely to initiate interactions than adults, and males more than females.

In all, juvenile males are slower to make the transition from juvenile to adult behaviour when their fathers are present. When fathers are absent, at about age 40 days juveniles start to wrestle less and box more, mount others, threaten each other, and begin to establish stable dominance hierarchies (Meany and Stewart 1981). When fathers are present, however, these changes are delayed: juvenile males continue to wrestle, they mount more but not as much as young rats without fathers, and stable dominance hierarchies haven't emerged even by 87 days.

The presence of fathers may be responsible: fathers engage in wrestling and boxing with their offspring (and consistently win these encounters), and by the end fathers are doing most of the chasing and mounting (Drews *et al.* 1982). Therefore, juvenile dominance and sexual behaviours may emerge more slowly when adult males are present and account for most of the dominance and sexual behaviours. Adult males also interact more with the juveniles than the adult females do, and may therefore play an important role in the socialisation of young rats.

Feeding and Drinking

How do Rats Choose What to Eat?

How rats choose what to eat is a large, well-studied area in animal behaviour. Rats are omnivores, and can potentially eat just about anything. This ability to eat anything makes them very flexible (unlike, say, the panda which eats mostly bamboo) and able to live in a huge variety of enviroments. Rats aren't "pre-programmed" to eat certain foods; each rat needs to figure out what to eat in its own environment.

However, one major danger of being able to eat anything is that there are a lot of poisonous foods out there, and a mistake can be fatal. How does a young rat decide which foods are good, and which are dangerous? Well, choosing the foods that other rats have eaten is a pretty good rule of thumb. Amazingly, this learning process happens at many different stages:

- *Fetal rats* detect odor-bearing chemicals that come from the mother's diet and cross the placental barrier. Shortly after birth, newborn rats will respond positively to those foods.
- *Nursing rats* receive information through their mother's milk about what their mother is eating, and prefer those foods.
- Later, when *young rats* are weaning and finding their own meals for the first time, they use adults of their colony as guides and forage where adults are eating, or at sites where adults have previously scent-marked.

- In *adolescence and adulthood*, when rats forage independently, their food choices can be influenced by social factors that occur in the home burrow far from feeding sites. The scent of foods recently eaten, carried on fur, whiskers, and breath of a forager can profoundly influence food choices in the rats that forager interacts with.

In addition, rats can learn what foods to avoid by experience. When a rat encounters a new food, he may sample just a small piece of it. If he becomes ill, he will associate the food with the illness and will avoid that food in the future (Garcia and Keolling 1966).

Eating and Drinking

Many rats cache their food. Caching makes a lot of adaptive sense, because food sources for the rats' wild ancestors may not have been constant, so securing and hiding food for later would be an important survival strategy. It is only in the captive environment that rats are fed an abundance of food regularly in the same location as the day before.

Rats eat by holding small items of food in their forepaws, and by eating large items straight from the ground.

Most domestic rats drink from water bottles, licking the water off the metal ball at the end of the waterspout. As they lick they briefly press the ball back into the spout with their tongue, which releases another drop of water that coats the metal ball. Rats may also lap standing water from a dish.

Pica: Eating of Non-nutritive Substances

Rats cannot vomit; they respond to nausea by eating non-nutritive substances (an activity called *pica*) like clay, dirt, and sometimes even bedding. Pica may be an adaptive response to nausea because nausea is frequently caused by a toxin: the non-nutritive substances (especially clay) may help dilute the toxin's effect on the body.

Chapter 8

Animal Sexual Behaviour

Animal sexual behaviour takes many different forms, even within the same species. Among animals other than humans, researchers have observed monogamy, promiscuity, sex between species, sexual arousal from objects or places, sex apparently via duress or coercion, copulation with dead animals, homosexual, heterosexual and bisexual sexual behaviour, situational sexual behaviour and a range of other practices. Related studies have noted diversity in sexed bodies and gendered behaviour, such as intersex and transgender animals.

The study of animal sexuality (and primate sexuality especially) is a rapidly developing field. It used to be believed that only humans and a handful of other species performed sexual acts other than for procreation, and that animals' sexuality was instinctive and a simple response to the "right" stimulation (sight, scent). Current understanding is that many species that were formerly believed monogamous have now been proven to be promiscuous or opportunistic in nature; a wide range of species appear both to masturbate and to use objects as tools to help them do so; in many species animals try to give and get sexual stimulation with others where procreation is not the aim; and homosexual behaviour has now been observed among 1,500 species and in 500 of those it is well documented.

Mating System

A mating system is a way in which a group is structured in relation to sexual behaviour. The precise meaning depends upon the context. With respect to higher animals, it specifies which males mate with which females, under which circumstances; recognised animal mating systems include monogamy, polygamy (which includes polygyny, polyandry, and polygynandry) and promiscuity. In plants, it refers to the degree and circumstances of outcrossing. In human sociobiology, the terms have been extended to encompass the formation of relationships such as marriage.

In Animals

The following are some of the mating systems generally recognised in animals:

- *Monogamy:* One male and one female have an exclusive mating relationship. The term "pair bonding" often implies this. It could also mean one or more animals have an exclusive relationship with one or more opposite-sex animals.
- *Polygamy:* Three types are recognised:
  - Polygyny (the most common polygamous mating system in vertebrates so far studied): One male has an exclusive relationship with two or more females
  - Polyandry: One female has an exclusive relationship with two or more males
  - *Polygynandry:* Two or more males have an exclusive relationship with two or more females; the numbers of males and females need not be equal, and in vertebrate species studied so far, the number of males is usually less.
- *Promiscuity:* A member of one sex within the social group mates with any member of the opposite sex.

These mating relationships may or may not be associated with social relationships, in which the sexual partners stay together to become parenting partners.

As the alternative term "pair bonding" implies, this is usual in monogamy. In many polyandrous systems, the males and the female stay together to rear the young.

In polygynous systems where the number of females paired with each male is low, the male will often stay with one female to help rear the young, while the other females rear their young on their own. In polygynandry, each of the males may assist one female; if all adults help rear all the young, the system is more usually called "communal breeding". In highly polygynous systems, and in promiscuous systems, paternal care of young is rare, or there may be no parental care at all.

It is important to realize that these descriptions are idealised, and that the social partnerships are often easier to observe than the mating relationships. In particular:

- the relationships are rarely exclusive for all individuals in a species. DNA fingerprinting studies have shown that even in pair-bonding, matings outside the pair (extra-pair copulations) occur with fair frequency, and a significant minority of offspring result from them.
- some species show different mating systems in different circumstances, for example in different parts of their geographical range, or under different conditions of food availability
- mixtures of the simple systems described above may occur.

Monogamous Pairing in Animals

Monogamous pairing in animals refers to the natural history of mating systems in which species pair bond to raise offspring. This is associated, usually implicitly, with sexual monogamy.

Animals

The evolution of mating systems in animals has received an enormous amount of attention from biologists. This section briefly reviews three main findings about the evolution of monogamy in animals. The amount of social monogamy in animals varies across taxa, with over 90% of birds engaging in social monogamy while only 7% of mammals were known to do the same.

This list is not complete. Other factors may also contribute to the evolution of social monogamy. Moreover, different sets of factors may explain the evolution of social monogamy in different species. There is no one-size-fits-all explanation of why different species evolved monogamous mating systems.

Sexual Dimorphism

Sexual dimorphism refers to differences in body characteristics between males and females. A frequently studied type of sexual dimorphism is body size. Males typically have larger bodies than females. In some species, however, females have larger bodies than males. Sexual dimorphism in body size has been linked to mating behaviour. In polygynous species, males compete for control over sexual access to females. Large males have an advantage in the competition for access to females, and they consequently pass their genes along to a greater number of offspring. This eventually leads to large differences in body size between males and females. Polygynous males are often 1.5 to 2.0 times larger in size than females. In monogamous species, on the other hand, males and females have more equal access to mates, so there is little or no sexual dimorphism in body size. From a new biological point of view, monogamy could result from mate guarding and is engaged as a result of sexual conflict following the hypothesis of mutual destruction.

Some researchers have attempted to infer the evolution of human mating systems from the evolution of sexual dimorphism. Several studies have reported a large amount of sexual dimorphism in *Australopithecus*, an evolutionary ancestor of human beings that lived between 2 and 5 million years ago. These studies raise the possibility that *Australopithecus* had a polygamous mating system. Sexual dimorphism then began to decrease. Studies suggest sexual dimorphism reached modern human levels around the time of *Homo Erectus* 0.5 to 2 million years ago. This line of reasoning suggests human ancestors started out polygamous and began the transition to monogamy somewhere between 0.5 million and 2 million years ago.

Attempts to infer the evolution of monogamy based on sexual dimorphism remain controversial for three reasons:

- The skeletal remains of *Australopithecus* are quite fragmentary. This makes it difficult to identify the sex of the fossils. Researchers sometimes identify the sex of the fossils by their size, which, of course, can exaggerate findings of sexual dimorphism.
- Recent studies using new methods of measurement suggest *Australopithecus* had the same amount of sexual dimorphism as modern humans. This raises questions about the amount of sexual dimorphism in *Australopithecus*.
- Humans may have been partially unique in that selection pressures for sexual dimorphism might have been related to the new niches that humans were entering at the time, and how that might have interacted with potential early cultures and tool use. If these early humans had a differentiation of gender roles, with men hunting and women gathering, selection pressures in favour of increased size may have been distributed unequally between the sexes.
- Even if future studies clearly establish sexual dimorphism in *Australopithecus*, other studies have shown the relationship between sexual dimorphism and mating system is unreliable. Some polygamous species show little or no sexual dimorphism. Some monogamous species show a large amount of sexual dimorphism.

Studies of sexual dimorphism raise the possibility that early human ancestors were polygamous rather than monogamous. But this line of research remains highly controversial. It may be that early human ancestors showed little sexual dimorphism, and it may be that sexual dimorphism in early human ancestors had no relationship to their mating systems.

Testis Size

The relative sizes of male testes often reflect mating systems. In species with promiscuous mating systems, where many males mate with many females, the testes tend to be relatively large. This appears to be the result of sperm competition. Males with large testes produce more sperm and thereby gain an advantage impregnating females. In polygynous species, where one male controls sexual access to females, the testes tend to be small. One male defends exclusive sexual access to a group of females and thereby eliminates sperm competition.

Studies of primates, including humans, support the relationship between testis size and mating system. Chimpanzees, which have a promiscuous mating system, have large testes compared to other primates. Gorillas, which have a polygynous mating system, have smaller testes than other primates. Humans, which have *a socially monogamous mating system, accompanied by moderate amounts of sexual non-monogamy*, have moderately sized testes. The moderate amounts of sexual non-monogamy in humans may result in a low to moderate amount of sperm competition. Also, notably, in the case of an avowedly sexually monogamous society, the occurrence of sexual nonmonogamy is typically culturally stigmatised, and therefore detecting its prevalence is inherently difficult, if indeed it is at all possible. At best, such statistics can be viewed as general approximations with a wide margin of error.

Although testis size in humans is consistent with the modern pattern of social monogamy accompanied by moderate sexual non-monogamy, this fact reveals little about when the modern pattern evolved. Did Homo Erectus have testes similar in size to modern humans? What about Australopithecus? It is not possible to measure the size of testes in the fossil remains of human ancestors. This limits the usefulness of testis size in understanding the evolution of monogamy in humans.

Monogamy as a Best Response

In species where the young are particularly vulnerable and may benefit from protection by both parents, monogamy may be an optimal strategy. The selection factors in favour of different mating strategies for a species of animal, however, may potentially operate on a large number of factors throughout that animal's life cycle. For instance, with many species of bear, the female will often drive a male off soon after mating, and will later guard her cubs from him similar to that of other animals after they are born. It is thought that this might be due to fact that too many bears close to one another may deplete an area of food resources for the relatively small but growing cubs.

Monogamy could be social but rarely genetic. Thierry Lodé argued that monogamy should result from conflict of interest between the sexes called sexual conflict. Organised from territory defence and mate guarding, monogamy appears as a response of male for the control of female sexuality, but exclusive monogamy would be rare and the biological evolution would privilege the diversity of sexual behaviours.

Sexual Conflict

Sexual conflict occurs when the two sexes have conflicting optimal fitness strategies concerning reproduction, leading to an evolutionary arms race between males and females. It has primarily been studied in animals, though it can in principle apply to any sexually reproducing organism, such as plants and fungi.

This can be in two forms:

1. Interlocus sexual conflict, where male alleles have conflicting interests with female alleles that are at a different genetic locus. This can be in the form of conflict over parental care: maternal alleles which cause the offspring to require a greater investment from the father, and paternal alleles which cause the offspring to require a greater investment from the mother. Another form is conflict over mating rate. Males frequently have a higher optimal mating rate than females (because in most animal species they invest fewer resources in offspring than females do), and so males have numerous adaptations to induce females to mate with them. Another well-documented example of interlocus sexual conflict is the seminal fluid of *Drosophila melanogaster*, which up-regulates females' egg-laying rate and reduces her desire to re-mate with another male (serving the male's interests), but also shortens the female's lifespan reducing her fitness.
2. Intralocus sexual conflict, where the same set of alleles in males and females have different optima. i.e. they are expressed differently in the sexes. A classic example is the human hip,

where females need larger hips for childbirth. The genes that affect hip size must reach a compromise that is at neither the male optimum nor the female optimum. In some cases, this conflict may be resolved through the differential expression of such loci in males and females, but evidence indicates that intralocus conflict may be an important constraint in the evolution of many traits.

Sexual conflict may lead to antagonistic co-evolution, in which one sex (usually males) evolves a favourable trait which is countered by a countering trait in the other sex. For example, male bean weevils (*Callosobruchus maculatus*) have spiny genitalia that are thought to allow them to copulate for a longer time without getting dislodged and hence transfer more sperm. However, this damages the female and reduces her fitness, so females have evolved the counter adaptation of kicking at males during mating, which reduces the time spent in copulation.

Some regard sexual conflict as a subset of sexual selection (which was traditionally regarded as mutualistic), while others suggest it is a separate evolutionary phenomenon.

Manifestations

There are a wide variety of manifestations of sexual conflict, occurring in a broad range of taxa, and showing extreme examples such as the well-known phenomenon of sexual cannibalism among insects, usually demonstrated by the female. One way of sorting these manifestations is by temporal relation to a given reference point, e.g. mating, fertilization, or birth.

Infanticide (Zoology)

In animals, infanticide involves the killing of young offspring by a mature animal of its own species, and is studied in zoology, specifically in the field of ethology. Ovicide is the analogous destruction of eggs. Although human infanticide has been widely studied, the practice has been observed in many other species throughout the animal kingdom. These include microscopic rotifers, insects, fish, amphibians, birds and mammals. Infanticide can be practiced by both males and females.

Infanticide caused by sexual conflict has the general theme of the killer (often male) becoming the new sexual partner of the victim's parent, which would otherwise be unavailable. This represents a gain in fitness by the killer, and a loss in fitness by the parents of the offspring killed. This is a type of evolutionary struggle between the two sexes, in which the victim sex may have counter-adaptations that reduce the success of this practice. It may also occur for other reasons, such as the struggle for food between females. In this case individuals may even kill closely related offspring.

Filial infanticide occurs when a parent kills its own offspring. This sometimes involves consumption of the young themselves, which is termed filial cannibalism. The behaviour is widespread in fishes, and is seen in terrestrial animals as well. Human infanticide has been recorded in almost every culture. A unique aspect of human infanticide is selective killing based on gender.

Background

Infanticide only came to be seen as a significant occurrence in nature quite recently. At the time it was first seriously treated by Yukimaru Sugiyama, infanticide was attributed to stress causing factors like overcrowding and captivity, and was considered pathological and maladaptive. Classical ethology held that conspecifics (members of the same species) rarely killed each other. By the 1980s it had gained much greater acceptance. Possible reasons it was not treated as a serious natural phenomenon include its abhorrence to people, the popular group and species selectionist notions of the time (the idea that individuals behave for the good of the group or species; compare with gene-centred view of evolution), and the fact that it is very difficult to observe in the field.

Infanticide Involving Sexual Conflict

This form of infanticide represents a struggle between the sexes, where one sex exploits the other, much to the latter's disadvantage. It is usually the male who benefits from this behaviour, though in cases where males play similar roles to females in parental care the victim and perpetrator may be reversed.

By Males

Hanuman langurs (or gray langurs) are an Old World monkey found in India. They are a social animal, living in groups. Each group is generally dominated by a single male, with many females, though the male must struggle for control of the group with other males. When a male tries to take over a group, there is a violent struggle with the existing male. If successful in overthrowing the previous male, infants of the females are then killed. This infanticidal period is limited to the window just after the group is taken over. Cannibalism, however, has not been observed in this species.

This behaviour not only reduces intraspecific competition between the incumbent's offspring and those of other males and increases the parental investment afforded to their own young, but also allows females to become sexually receptive sooner. This is because females of this species, as well as many other mammals, do not ovulate during the period in which they produce milk. It then becomes easier to see how this behaviour could have evolved. If a male kills a female's young, they stop lactating and are able to become pregnant again. As males are in a constant struggle to protect their group, those that express infanticidal behaviour will contribute a larger portion to future gene pools.

Similar behaviour is also seen in male lions, among other species, who also kill young cubs, allowing them to impregnate the females. Unlike langurs, male lions live in small groups, which cooperate to take control of a pride from an existing group. They will attempt to kill any cubs that are roughly 9 months old or less, though as in other species, the female will attempt to defend their cubs viciously. Males have, on average, only a two year window in which to pass on their genes, and female lions only give birth once every two years, so the selective pressure from them to behave like this is strong. In fact it is estimated that a quarter of cubs dying in the first year of their life are victims of infanticide.

Male mice show great variation in behaviour over time. After fertilizing a female, they become aggressive towards mouse pups for three weeks, killing any they come across. After this period however, their behaviour changes dramatically, and they become paternal, caring for their own offspring. This lasts for almost two months, but afterwards they become infanticidal once more.

It is no coincidence here that the female gestation period is three weeks as well, or that it takes roughly two months for pups to become fully weaned and leave their nest. The proximate mechanism that allows for the correct timing of these periods involves circadian rhythms, each day and night cycle affecting the mouse's internal neural physiology, and disturbances in the duration of these cycles results in different periods of time between behaviours. The adaptive value of this behaviour switching is twofold; infanticide removes competitors for when the mouse does have offspring, and allows the female victims to be impregnated earlier than if they continued to care for their young, as mentioned above.

Gerbils, on the other hand, no longer commit infanticide once they have paired with a female, but actively kill and eat other offspring when young. The females of this species behave much like male mice, hunting down other litters except when rearing their own.

Prospective Infanticide

Prospective infanticide is a subset of sexual competition infanticide in which young born *after* the arrival of the new male are killed. This is less common than infanticide of existing young, but can still increase fitness in cases where the offspring could not possibly have been fathered by the new mate, i.e. one gestation or fertility period. This is known to occur in lions and langurs, and has also been observed in other species such as house wrens. In birds, however, the situation is more complex, as female eggs are fertilized one at a time, with a 24-hour delay between each. Males may destroy clutches laid 12 days or more after their arrival, though their investment of around 60 days of parental care is large, so a high level of parental certainty is needed.

By Females

Females are also known to display infanticidal behaviour. This may appear unexpected, as the conditions described above do not apply. Males are not always an unlimited resource though - in some species, males provide parental care to their offspring, and females may compete indirectly with others by killing their offspring, freeing up the limiting resource that the males represent. This has been documented in research by Stephen Emlen and Natalie Demong on wattled jacanas (*Jacana jacana*), a tropical wading bird. With the wattled jacana, it is exclusively the male sex that broods, while females defend their territory. In this experiment Demong and Emlen found that removing females from a territory resulted in nearby females attacking the chicks of the male in most cases, evicting them from their nest. The males then fertilized the offending females and cared for their young. Emlen describes how he "shot a female one night, and [...] by first light a new female was already on the turf. I saw terrible things—pecking and picking up and throwing down chicks until they were dead. Within hours she was soliciting the male, and he was mounting her the same day. The next night I shot the other female, then came out the next morning and saw the whole thing again."

Infanticide is also seen in giant water bugs. *Lethocerus deyrollei* is a large and nocturnal predatory insect found in still waters near vegetation. In this species the males take care of masses of eggs by keeping them hydrated with water from their bodies. Without a male caring for the eggs like this, they become desiccated and will not hatch. In this species, males are a scarce resource that females must sometimes compete for. Those that cannot find a free male often stab the eggs of a brooding one. As in the above case, males then fertilize this female and care for her eggs.

Noritaka Ichikawa has found that males only moisten their eggs during the first 90 seconds or so, after which all of the moisture on their bodies has evaporated. However, they guard the egg masses for as long as several hours at a time, when they could be hunting prey. They do not seem to prevent further evaporation by staying guard, as males that only guarded the nest for short periods were seen to have similar hatching rates in a controlled experiment where there were no females present. It seems rather that males are more successful in avoiding infanticidal females when they are out of the water with their eggs, which might well explain the ultimate cause of this behaviour.

Resource Competition

Black-tailed prairie dogs are colonially-living, harem-polygynous squirrels found mainly in the United States. Their living arrangement involves one male living with four or so females in a territory defended by all individuals, and underground nesting. Black-tails only have one litter per year, and are in estrous for only a single day around the beginning of spring.

A seven-year natural experiment by John Hoogland and others from the University of Princeton revealed that infanticide is widespread in this species, including infanticide from invading males and immigrant females, as well as occasional cannibalism of an individual's own offspring.

The surprising finding of the study was that by far the most common type of infanticide involved the killing of close kin's offspring. This seems illogical, as kin selection favours behaviours that promote the well-being of closely related individuals. It was postulated that this form of infanticide is more successful than trying to kill young in nearby groups, as the whole group must be bypassed in this case, while within a group only the mother need be evaded. Marauding behaviour is evidently adaptive, as infanticidal females had more and healthier young than others, and were heavier themselves as well. This behaviour appears to reduce competition with other females for food, and future competition among offspring.

Similar behaviour has been reported in the meerkat (*Suricata suricatta*), including cases of females killing their mother's, sister's, and daughter's offspring. Infanticidal raids from neighbouring groups also occurred.

Other

Bottlenose dolphins have been reported to kill their young through impact injuries. Dominant male langurs tend to kill the existing young upon taking control of a harem. There has been sightings of infanticide in the leopard population.

Costs and Defences

Costs of the Behaviour: While it may be beneficial for some species to behave this way, infanticide is not without risks to the perpetrator. Having already exhausted energy and perhaps sustained serious wounds in a fight with another male, attacks from females who vigorously defend their offspring may be telling for harem-polygynous males, with a risk of infection. It is also energetically costly to pursue a mother's young, which may try to escape.

Costs of the behaviour described in prairie dogs include the risk to an individual of losing their own young while killing another's, not to mention the fact that they are killing their own relatives. In a species where infanticide is common, perpetrators may well be victims themselves in the future, such that they come out no better off; but as long as an infanticidal individual gains in reproductive output by its behaviour, it will tend to become common. Further costs of the behaviour in general may be induced by counter-strategies evolved in the other sex, as described below.

As a Cost of Social Behaviour

Taking a broader view of the black-tailed prairie dog situation, infanticide can be seen as a cost of social living. If each female were to have her own private nest away from others, she would be much less likely to have her infants killed when absent. This, and other costs such as increased spread of parasites, must be made up for by other benefits, such as group territory defence and increased awareness of predators. An avian example published in *Nature* is acorn woodpeckers. Females nest together, possibly because those nesting alone have their eggs constantly destroyed by rivals. Even so, eggs are consistently removed at first by nest partners themselves, until the entire group lays on the same day. They then cooperate and incubate the eggs as a group, but by this time a significant proportion of their eggs have been lost because of this ovicidal behaviour.

Counter-strategies

Because this form of infanticide reduces the fitness of killed individuals' parents, animals have evolved a range of counter-strategies against this behaviour. These may be divided into two very different classes - those that tend to prevent infanticide, and those that minimise losses.

Loss Minimisation

Some females abort or resorb their own young while they are still in development after a new male takes over; this is known as the Bruce effect. This may prevent their young from being killed after birth, saving the mother wasted time and energy. However, this strategy also benefits the new male. In mice this can occur by the proximate mechanism of the female smelling the odor of the new male's urine.

Preventative Adaptations

Infanticide in burying beetles may have led to male parental care. In this species males often cooperate with the female in preparing a piece of carrion, which is buried with the eggs and eaten by the larvae when they hatch. Males may also guard the site alongside the female. It is apparent from experiments that this behaviour does not provide their young with any better nourishment, nor is it of any use in defending against predators. However, other burying bugs may try to take their nesting space. When this occurs, a male-female pair is over twice as successful in nest defence, preventing the ovicide of their offspring.

Female langurs may leave the group with their young alongside the outgoing male, and others may develop a *false estrous* and allow the male to copulate, deceiving him into thinking she is actually sexually receptive. Females may also have sexual liaisons with other males. This promiscuous behaviour is adaptive, because males will not know whether it is their own offspring they are killing or not, and may be more reluctant or invest less effort in infanticide attempts. Lionesses cooperatively guard against scouting males, and a pair were seen to violently attack a male after it killed one of their young. Resistance to infanticide is also costly, though: for instance, a female may sustain serious injuries in defending her young. At times it is simply more advantageous to submit than to fight.

Infanticide by Parents and Caregivers

Filial infanticide occurs when a parent kills its own offspring. Both male and female parents have been observed to do this, as well as sterile worker castes in some eusocial animals.

Maternal

Maternal infanticide occurs when newborn offspring are killed by their mother. This is sometimes seen in pigs, a behaviour known as savaging, which affects up to 5% of gilts. Similar behaviour has been observed in various animals such as rabbits and burying beetles.

Paternal

Paternal infanticide—where fathers eat their own offspring—may also occur. When young bass hatch from the spawn, the father guards the area, circling around them and keeping them together, as well as providing protection from would-be predators. After a few days, most of the fish will swim away. At this point the male's behaviour changes: instead of defending the stragglers, he treats them as any other small prey, and eats them.

Worker Caste Killing Young

Honey bees may become infected with a bacterial disease called foul brood, which attacks the developing bee larva while still living in the cell. Some hives however have evolved a behavioural adaptation that resists this disease: the worker bees selectively kill the infected individuals by removing them from their cells and tossing them out of the hive, preventing it from spreading.

The genetics of this behaviour are quite complex. Experiments by Rothenbuhler showed that the 'hygienic' behaviour of the queen was lost by crossing with a non-hygienic drone. This means that the trait must be recessive, only being expressed when both alleles contain the gene for hygienic behaviour. Furthermore, the behaviour is dependent on two separate loci. A backcross produced a mixed result. The hives of some offspring were hygienic, while others were not. There was also a third type of hive where workers removed the wax cap of the infected cells, but did nothing more. What was not apparent was the presence of a fourth group who threw diseased larvae out of the hive, but did not have the uncapping gene. This was suspected by Rothenbuhler however, who manually removed the caps, and found some hives proceeded to clear out infected cells.

Humans and Infanticide

Perceptions and Importance: Like many of the darker aspects of nature, infanticide is a subject that many find discomforting. Cornell University ethologist Glenn Hausfater states that "infanticide has not received much study because it's a repulsive subject [...] Many people regard it as reprehensible to even think about it."

Research into infanticide in animals is in part motivated by the desire to understand human behaviours, such as child abuse. Hausfater explains that researchers are "trying to see if there's any connection between animal infanticide and child abuse, neglect and killing by humans. We just don't know yet what the connections are."

Infanticide by Humans

Infanticide has been, and still is, practiced by some cultures, groups, or individuals. It is often, but not always, the mother who commits the act. In many past societies, certain forms of infanticide were considered permissible, whereas in most modern societies the practice is considered immoral and criminal. Nonetheless, it still takes place — in the Western world usually because of the parent's mental illness or violent behaviour, and in some poor countries as a form of population control, sometimes with tacit societal acceptance. Female infanticide, a form of sex-selective infanticide, is more common than the killing of male offspring.

During pregnancy, there is a two-way traffic of immunologically active cell lines through the placenta. Fetal lymphocyte lines may survive in women even decades after giving birth. These cells may serve the adaptive interest of the mother, however, they may also serve conflicting interests of the fetus or those of the father. This mixture of shared or contradicting interests have been hypothesized to give rise to diseases like autoimmune diseases, infertility, and habitual spontaneous abortion (i.e., early miscarriage) in humans.

Harem (Zoology)

The term harem is used in zoology to describe the social organisation of certain herbivore species, such as those in the Hominidae and Equidae families, into groups of females and young surrounding a single dominant male. Non dominant males may form bachelor groups, or may occasionally be permitted to remain on the periphery of the harem.

Social Monogamy

Social monogamy refers to a male and female's social living arrangement (e.g., shared use of a territory, behaviour indicative of a social pair, and/or proximity between a male and female) without inferring any sexual interactions or reproductive patterns. In humans, social monogamy equals monogamous marriage. Sexual monogamy is defined as an exclusive sexual relationship between a female and a male based on observations of sexual interactions. Finally, the term genetic monogamy is used when DNA analyses can confirm that a female-male pair reproduce exclusively with each other. A combination of terms indicates examples where levels of relationships coincide, e.g., sociosexual and sociogenetic monogamy describe corresponding social and sexual, and social and genetic monogamous relationships, respectively. —Reichard, 2003, p. 4

Whatever makes a pair of animals socially monogamous does not necessarily make them sexually or genetically monogamous. Social monogamy, sexual monogamy, and genetic monogamy can occur in different combinations.

Social monogamy is relatively rare in the animal kingdom. The actual incidence of social monogamy varies greatly across different branches of the evolutionary tree. Over 90 percent of avian species are socially monogamous.

This stands in contrast to mammals. Only 3 percent of mammalian species are socially monogamous, although up to 15 percent of primate species are socially monogamous. Social monogamy has also been observed in reptiles, fish, and insects.

Sexual monogamy is also rare among animals. Many socially monogamous species engage in extra-pair copulations, making them sexually non-monogamous. For example, while over 90% of birds are socially monogamous, "on average, 30 percent or more of the baby birds in any nest [are] sired by someone other than the resident male." Patricia Adair Gowaty has estimated that, out of 180 different species of socially monogamous songbirds, only 10% are sexually monogamous.

The incidence of genetic monogamy, determined by DNA fingerprinting, varies widely across species. For a few rare species, the incidence of genetic monogamy is 100 percent, with all offspring genetically related to the socially monogamous pair. But genetic monogamy is strikingly low in other species. Barash and Lipton note:

The highest known frequency of extra-pair copulations are found among the fairy-wrens, lovely tropical creatures technically known as *Malurus splendens* and *Malurus cyaneus*. More than 65 percent of all fairy-wren chicks are fathered by males outside the supposed breeding group. —Barash & Lipton, 2001, p. 12

Such low levels of genetic monogamy have surprised biologists and zoologists, forcing them to rethink the role of social monogamy in evolution. They can no longer assume social monogamy determines how genes are distributed in a species. The lower the rates of genetic monogamy among socially monogamous pairs, the less of a role social monogamy plays in determining how genes are distributed among offspring.

Polygamy

Polygamy is defined as a mating structure in which a single individual of one gender has exclusive access to several individuals of the opposite gender. It takes two main forms – polygyny and polyandry. As polygyny is the most common form of polygamy among vertebrates (including humans, to some extent), it has been studied far more extensively than polyandry.

Polygyny

In some species, notably those with harem-like structures, only one of a few males in a group of females will mate. Technically, polygyny in sociobiology and zoology is defined as a system in which a male has a relationships with more than one female, but the females are predominantly bonded to a single male. Should the active male be driven out, killed, or otherwise removed from the group, in a number of species the new male will ensure that breeding resources are not wasted on another male's young. The new male may achieve this in many different ways, including:

- *Competitive Infanticide:* In lions, hippopotamuses, and some monkeys, the new male will kill the offspring of the previous alpha male to cause their mothers to become receptive to his sexual advances since they are no longer nursing.
- *Harassment to Miscarriage:* Amongst wild horses and baboons, the male will "systematically harass" pregnant females until they miscarry.
- *Pheromone based Spontaneous Abortion:* In some rodents such as mice, a new male with a different scent will cause females who are pregnant to spontaneously fail to implant recently fertilized eggs. This does not require contact; it is mediated by scent alone. It is known as the Bruce-Parkes effect.

Promiscuity

Two examples of systems in primates are promiscuous mating chimpanzees and bonobos. These species live in social groups consisting of several males and several females. Each female copulates with many males, and vice versa. In bonobos, the amount of promiscuity is particularly striking because bonobos use sex to alleviate social conflict as well as to reproduce.

Seasonal Breeder

Seasonal breeders are animal species that successfully mate only during certain times of the year. These times of year allow for the births at a time optimal for the survival of the young in terms of factors such as temperature, food and water. Related sexual interest and behaviours are expressed and accepted only during this period.

Female seasonal breeders will have one or more estrus cycles only when she is "in season" or fertile and receptive to mating. At other times of the year, they will be anestrus.

Similarly, male seasonal breeders may exhibit changes in testosterone levels, testes weight and fertility depending on the time of year.

Seasonal breeders are distinct from opportunistic breeders, which mate whenever the conditions of their environment become favourable, and continuous breeders like humans that mate year-round.

Physiology

The hypothalamus is considered to be the central control for reproduction. Hence, factors that determine when a seasonal breeder will be ready for mating affect this tissue.

This is achieved specifically through changes in the production of the hormone GnRH. GnRH in turn transits to the pituitary where it promotes the secretion into the bloodstream of the gonadotropin LH, a pituitary hormone critical for reproductive function and behaviour. Changes in gonadotropin secretion initiate the end of anestrus in females.

Factors that Determine Time of Fertility

Photoperiod: When a seasonal breeder is ready for mating is strongly regulated by length of day (photoperiod) and thus season. Photoperiod likely affects the seasonal breeder through changes in melatonin secretion by the pineal gland that ultimately alter GnRH release by the hypothalamus.

Hence, seasonal breeders can be divided into groups based on when they are fertile. "Long day" breeders cycle when days get longer (spring) and are anestrus in fall and winter. "Short day" breeders cycle when the length of daylight shortens (fall) and are anestrus in spring and summer. Domestication has allowed cattle and swine to be liberated from breeding seasonality. Day length variations with latitude can also impact breeding. For instance, sheep and goats in tropical climes may breed throughout the year while those in more polar arctic areas may have a shortened season.

Females are generally more sensitive to changes in day length. For instance, unlike mares, stallions remain fertile year-round, suffering only some declines in sexual behaviour and sperm production out of season.

Other Factors

Other factors that affect breeding time include the presence of a ready and available mate. For instance, the presence of a fertile male will induce an estrus cycle in a doe shortly after introduction.

Further environmental factors can include nutrition, chemosensory and hormonal cues. Weight and age are other factors.

Partial List of Seasonal Breeders

Many non-mammals are seasonal breeders, such as many birds and fish. Here is partially listed those that are mammals.

Long Day Breeders

- Ring-tailed lemur
- Horse
- Hamster
- Groundhog
- Mink.

Short Day Breeders

- Sheep
- Goat
- Fox
- Deer, Red Deer
- Elk
- Moose.

Summer Breeders

- Ruffed lemur (May - July)
- Select species of hamster, vole and mouse.

Interpretation Bias

The field of study of sexuality in non-human species has been a long standing taboo, with researchers either failing to observe or miscategorising and misdescribing sexual behaviour which does not meet their preconceptions.

More current research provides views such as that of the Natural History Museum at the University of Oslo, which in 2006 held an exhibition on animal sexuality:

Many researchers have described homosexuality as something altogether different from sex. They must realise that animals can have sex with who they will, when they will and without consideration to a researcher's ethical principles.

An example of overlooking behaviour relates to descriptions of giraffe mating:

When nine out of ten pairings occur between males, "[e]very male that sniffed a female was reported as sex, while anal intercourse with orgasm between males was only [categorised as] 'revolving around' dominance, competition or greetings.

Sex for Pleasure

It is a common myth that animals do not (as a rule) have sex for pleasure, or alternatively that humans, pigs (and perhaps dolphins and one or two species of primate) are the only species which do. This is sometimes formulated "animals mate only for reproduction".

Science cannot conclusively say at present what animals do or do not find "pleasurable", a question considered in more depth under Emotion in animals. The urban considers this particular view in depth. Its conclusions are broadly that the statement is true, but only using a *very specific definition* of "sex for pleasure", in which sexual acts tied to a reproductive cycle or for which an alternative explanation can be asserted, are ignored, as is all sexual activity that does not involve penetration. Animals put themselves at risk to engage in sex, and as a result, most species have evolved sexual signals (usually scent and behaviour) to indicate the presence of receptive periods. During these, sex is sought, and outside these it is usually not sought (or is sought but not permitted). Snopes comments that this is not in fact a reflection of whether sex is pleasurable or not, but rather a reflection of whether individuals have sex at arbitrary times. They conclude:

> *"Of course, we have to make many seemingly artificial distinctions to arrive at our conclusion. Animals other than humans have no awareness that their sexual activities are connected with reproduction: They engage in sex because they're biologically driven to do so, and if the fulfillment of their urges produces a physical sensation we might appropriately call 'pleasure,' it isn't the least bit affected by the possibility (or impossibility) of producing offspring. We are also discounting cases in which animals do engage in sex even though reproduction is an impossibility because we claim there are other*

> *'purposes' (of which the animals themselves are unaware) at play. (For example, the females of some species of birds will invite males to mate with them even after they have laid their eggs, but we ascribe a purpose to this behaviour: this is a biological "trick" to fool males into caring for hatchlings they didn't father.) We also employ subjective terms such as 'willingly' and 'regularly' in claiming that bonobos and dolphins are the only other animals who "willingly (and regularly) engage in sex with each other" ... and even then it may be the case that these species have some other 'purpose' for doing so that we haven't yet discovered..."*

A 2006 Danish Animal Ethics Council report which examined current knowledge of animal sexuality in the context of legal queries concerning sexual acts by humans, has the following comments, primarily related to domestically common animals:

Even though the evolution-related purpose of mating can be said to be reproduction, it is not actually the creating of offspring which originally causes them to mate. It is probable that they mate because they are motivated for the actual copulation, and because this is connected with a positive experience.

It is therefore reasonable to assume that there is some form of pleasure or satisfaction connected with the act. This assumption is confirmed by the behaviour of males, who in the case of many species are prepared to work to get access to female animals, especially if the female animal is in oestrus, and males who for breeding purposes are used to having sperm collected become very eager, when the equipment they associate with the collection is taken out.

There is nothing in female mammals' anatomy or physiology, that contradicts that stimulation of the sexual organs and mating is able to be a positive experience. For instance, the clitoris acts in the same way as with women, and scientific studies have shown that the success of reproduction is improved by stimulation of clitoris on (among other species) cows and mares in connection with insemination, because it improves the transportation of the sperm due to contractions of the inner genitalia. This probably also concerns female animals of other animal species, and contractions in the inner genitals are seen e.g. also during orgasm for women. It is therefore reasonable to assume that sexual intercourse may be linked with a positive experience for female animals.

Types of Activity

Autoeroticism or Masturbation: It appears that many animals, both male and female, masturbate, both when partners are available and otherwise.

For example, Petplace comments in its guide on assessing potential breeding stock purchases: *"Masturbation is a normal behaviour in all stallions that does not reduce semen production or performance in the breeding shed"* Likewise a review from the University of Pennsylvania School of Veterinary Medicine says:

> *the behaviour known within the horse breeding industry as* masturbation. *This involves normal periodic erections and penile movements. This behaviour, both from the descriptive field studies cited above and in extensive study of domestic horses, is now understood as normal, frequent behaviour of male equids. Attempting to inhibit or punish masturbation, for example by tying a brush to the area of the flank underside where the penis rubs into contact with the underside, which is still a common practice of horse managers regionally around the world, often leads to increased masturbation and disturbances of normal breeding behaviour.*
>
> —Sue M. McDonnell, *Sexual Behaviour – Current Topics in Applied Ethology and Clinical Methods*

Castration does not prevent masturbation, as it is observed in geldings. Masturbation is common in both mares and stallions, before and after puberty. Sexologist Havelock Ellis in his 1927 "Studies in the Psychology of Sex" identified bulls, goats, sheep, camels and elephants as species known to practice autoeroticism, adding of some other species:

> *I am informed by a gentleman who is a recognised authority on goats, that they sometimes take the penis into the mouth and produce actual orgasm, thus practicing auto-fellatio. As regards ferrets ... "if the bitch, when in heat, cannot obtain a dog [ie, male ferret] she pines and becomes ill. If a smooth pebble is introduced into the hutch, she will masturbate upon it, thus preserving her normal health for one season. But if this artificial substitute is given to her a second season, she will not, as formerly, be content with it." Blumenbach observed a bear act somewhat similarly on seeing other*

> *bears coupling, and hyenas, according to Ploss and Bartels, have been seen practicing mutual masturbation by licking each other's genitals.*

In his 1999 book, *Biological exuberance*, Bruce Bagemihl PhD documents (p. 71, 209–210) that:

> *Autoeroticism also occurs widely among animals, both male and female. A variety of creative techniques are used, including genital stimulation using the hand or front paw (primates, Lions), foot (Vampire Bats, primates), flipper (Walruses), or tail (Savanna Baboons), sometimes accompanied by stimulation of the nipples (Rhesus Macaques, Bonobos); auto-fellating or licking, sucking and/or nuzzling by a male of his own penis (Common Chimpanzees, Savanna Bonobos, Vervet Monkeys, Squirrel Monkeys, Thinhorn Sheep, Bharal, Aovdad, Dwarf Cavies); stimulation of the penis by flipping or rubbing it against the belly or in its own sheath (White-tailed and Mule Deer, Zebras and Takhi); spontaneous ejaculations (Mountain Sheep, Warthogs, Spotted Hyenas); and stimulation of the genitals using inanimate objects (found in several primates and cetaceans).*

Many birds masturbate by mounting and copulating with tufts of grass, leaves or mounds of earth, and some mammals such as primates and dolphins also rub their genitals against the ground or other surfaces to stimulate themselves.

Autoeroticism in female mammals, as well as heterosexual and homosexual intercourse (especially in primates), often involves direct or indirect stimulation of the clitoris. This organ is present in the females of all mammalian species and several other animal groups.

And that:

> *Apes and Monkeys use a variety of objects to masturbate with and even deliberately create implements for sexual stimulation [...] often in highly creative ways.*

Petter Bøckman of the Natural History Museum at the University of Oslo commented (in respect of a 2006 exhibition on homosexuality in the animal kingdom) that:

> *Masturbation is common in the animal kingdom ... We have a Darwinist mentality that all animals only have sex to procreate. But there are plenty of animals who*

will masturbate when they have nothing better to do. Masturbation has been observed among primates, deer, killer whales and penguins, and we're talking about both males and females. They rub themselves against stones and roots. Orangutans are especially inventive. They make dildos of wood and bark.

Oral Sex

Animals of several species are documented as engaging in both autofellatio and oral sex. Although easily confused by lay-people, this is a separate and sexually oriented behaviour, distinct from non-sexual grooming or the investigation of scents.

Auto-fellatio or oral sex in animals is documented in goats, primates, hyaenas, bats and sheep.

Contraceptive Sex

Among monkeys, Lionel Tiger and Robin Fox conducted a study on how Depo-Provera contraceptives lead to decreased male attractiveness to females and eventually to male homosexuality. Janet E. Smith summarizes the findings as follows:

The study in the early 70s involved a tribe of monkeys. The alpha monkey of this tribe, named Austin, chose three female monkeys to be his exclusive sexual partners. Austin had a grand time with these three female monkeys. Then the researchers injected Austin's three females with the contraceptive Depo-Provera. Austin stopped having sex with them and chose other female monkeys to be his sexual partners. Then they contracepted all of the females in the tribe. The males stopped having sex with the females and started behaving in a turbulent and confused manner.

Homosexual Behaviour in Animals

Homosexual behaviour in animals refers to the documented evidence of homosexual and bisexual behaviour in non-human species. Such behaviours include sex, courtship, affection, pair bonding, and parenting among same sex animals. A 1999 review by researcher Bruce Bagemihl shows that homosexual behaviour has been observed in close to 1,500 species, ranging from primates to gut worms, and is well documented for 500 of them.

Animal sexual behaviour takes many different forms, even within the same species. The motivations for and implications of these behaviours have yet to be fully understood, since most species have yet to be fully studied. According to Bagemihl, "the animal kingdom does it with much greater sexual diversity — including homosexual, bisexual and nonreproductive sex — than the scientific community and society at large have previously been willing to accept." Current research indicates that various forms of same-sex sexual behaviour are found throughout the animal kingdom. A new review made in 2009 of existing research showed that same-sex behaviour is a nearly universal phenomenon in the animal kingdom, common across species.

Homosexual behaviour is best known from social species. According to geneticist Simon Levay in 1996, "Although homosexual behaviour is very common in the animal world, it seems to be very uncommon that individual animals have a long-lasting predisposition to engage in such behaviour to the exclusion of heterosexual activities. Thus, a homosexual orientation, if one can speak of such thing in animals, seems to be a rarity. One species in which exclusive homosexual orientation occurs, however, is that of domesticated sheep (*Ovis aries*). "About 10% of rams (males) refuse to mate with ewes (females) but do readily mate with other rams."

The observation of homosexual behaviour in animals can be seen as both an argument for and against the acceptance of homosexuality in humans, and has been used especially against the claim that it is a *peccatum contra naturam* ('sin against nature'). For instance, homosexuality in animals was cited in the United States Supreme Court's decision in *Lawrence v. Texas* which struck down the sodomy laws of 14 states. Whether animal sexuality has logical, ethical, or moral implications in human sexuality is also a source of debate.

Applying the Term "Homosexual" to Animals

The term *homosexual* was coined by Karl-Maria Kertbeny in 1868 to describe same-sex sexual attraction and sexual behaviour in humans. Its use in animal studies has been controversial for two main reasons: animal sexuality and motivating factors have been and remain poorly understood, and the term has strong cultural implications in western society that are irrelevant for species other than humans. Thus homosexual behaviour has been given a number of terms over the years. When describing animals, the term "homosexual" is preferred over "gay", "lesbian" and other terms currently in use, as these are seen as even more bound to human homosexuality.

Animal preference and motivation is always inferred from behaviour. In wild animals, researchers will as a rule not be able to map the entire life of an individual, and must infer from frequency of single observations of behaviour. The correct usage of the term *homosexual* is that an animal *exhibits homosexual behaviour* or even *same-sex sexual behaviour*; however, this article conforms to the usage by modern research applying the term *homosexuality* to all sexual behaviour (copulation, genital stimulation, mating games and sexual display behaviour) between animals of the same sex. In most instances, it is presumed that the homosexual behaviour is but part of the animal's overall sexual behavioural repertoire, making the animal "bisexual" rather than "homosexual" as the terms are commonly understood in humans, but cases of homosexual preference and exclusive homosexual pairs are known.

Research on Homosexual Behaviour in Animals

The presence of same-sex sexual behaviour was not 'officially' observed on a large scale until recent times, possibly due to observer bias caused by social attitudes to same-sex sexual behaviour, innocent confusion, or even from a fear of "being ridiculed by their colleagues." Georgetown University biologist Janet Mann states "Scientists who study the topic are often accused of trying to forward an agenda, and their work can come under greater scrutiny than that of their colleagues who study other topics." They also noted "Not every sexual act has a reproductive function ... that's true of humans and non-humans." It appears to be widespread amongst social birds and mammals, particularly the sea mammals and the primates. The true extent of homosexuality in animals is not known.

While studies have demonstrated homosexual behaviour in a number of species, Petter Bøckman, the scientific advisor of the exhibition Against Nature? In 2007, speculated that the true extent of the phenomenon may be much larger than was then recognised:

No species has been found in which homosexual behaviour has *not* been shown to exist, with the exception of species that never have sex at all, such as sea urchins and aphis. Moreover, a part of the animal kingdom is hermaphroditic, truly bisexual. For them, homosexuality is not an issue.

An example of overlooking homosexual behaviour is noted by Bruce Bagemihl describing mating giraffes where nine out of ten pairings occur between males.

Every male that sniffed a female was reported as sex, while anal intercourse with orgasm between males was only "revolving around" dominance, competition or greetings.

Some researchers believe this behaviour to have its origin in male social organisation and social dominance, similar to the dominance traits shown in prison sexuality. Others, particularly Joan Roughgarden, Bruce Bagemihl, Thierry Lodé and Paul Vasey suggest the social function of sex (both homosexual and heterosexual) is not necessarily connected to dominance, but serves to strengthen alliances and social ties within a flock. Others have argued that social organisation theory is inadequate because it cannot account for some homosexual behaviours, for example, penguin species where same-sex individuals mate for life and refuse to pair with females when given the chance. While reports on many such mating scenarios are still only anecdotal, a growing body of scientific work confirms that permanent homosexuality occurs not only in species with permanent pair bonds, but also in non-monogamous species like sheep.

One report on sheep cited below states:

Approximately 8% of rams exhibit sexual preferences, that is even when given a choice for male partners (male-oriented rams) in contrast to most rams, which prefer female partners (female-oriented rams). We identified a cell group within the medial preoptic area/anterior hypothalamus of age-matched adult sheep that was significantly larger in adult rams than in ewes.

In fact, apparent homosexual individuals are known from all of the traditional domestic species, from sheep, cattle and horses to cats, dogs and budgerigars.

Genetic and Physiological Basis for Homosexual Animal Behaviour

Researchers found that disabling the (fucose mutarotase) FucM gene in laboratory mice – which influences the levels of estrogen to which the brain is exposed – caused the female mice to behave as if they were male as they grew up. "The mutant female mouse underwent a slightly altered developmental programme in the brain to resemble the male brain in terms of sexual preference" said Professor Chankyu Park of the Korea Advanced Institute of Science and Technology in Daejon, South Korea, who led the research. His most recent findings have been published in the BMC Genetics journal on July 7, 2010.

In March 2011, research shows that serotonin is involved in the mechanism of sexual orientation of mice. Some selected species and groups

Birds

Black Swans: An estimated one-quarter of all black swans pairings are homosexual and they steal nests, or form temporary threesomes with females to obtain eggs, driving away the female after she lays the eggs. More of their cygnets survive to adulthood than those of different-sex pairs, possibly due to their superior ability to defend large portions of land. The same reasoning has been applied to male flamingo pairs raising chicks.

Gulls: Studies have shown that 10 to 15 percent of female western gulls in some populations in the wild exhibit homosexual behaviour.

Ibises: Research has shown that the environmental pollutant methylmercury can increase the prevalence of homosexual behaviour in male American White Ibis. The study involved exposing chicks in varying dosages to the chemical and measuring the degree of homosexual behaviour in adulthood. The results discovered was that as the dosage was increased the likelihood of homosexual behaviour also increased. The endocrine blocking feature of mercury has been suggested as a possible cause of sexual disruption in other bird species.

Mallards: Mallards form male-female pairs only until the female lays eggs, at which time the male leaves the female. Mallards have rates of male-male sexual activity that are unusually high for birds, in some cases, as high as 19% of all pairs in a population.

Penguins: In early February 2004 the *New York Times* reported that Roy and Silo, a male pair of chinstrap penguins in the Central Park Zoo in New York City had successfully hatched and fostered a female chick from a fertile egg they had been given to incubate. Other penguins in New York zoos have also been reported to have formed same-sex pairs.

Zoos in Japan and Germany have also documented homosexual male penguin couples. The couples have been shown to build nests together and use a stone as a substitute for an egg. Researchers at Rikkyo University in Tokyo found 20 homosexual pairs at 16 major aquariums and zoos in Japan.

Bremerhaven Zoo in Germany attempted to encourage reproduction of endangered Humbolt penguins by importing females from Sweden and separating three male pairs, but this was unsuccessful.

The zoo's director said that the relationships were "too strong" between the homosexual pairs. German gay groups protested at this attempt to break up the male-male pairs but the zoo's director was reported as saying "We don't know whether the three male pairs are really homosexual or whether they have just bonded because of a shortage of females... nobody here wants to forcibly separate homosexual couples."

A pair of male Magellanic penguins who had shared a burrow for six years at the San Francisco Zoo and raised a surrogate chick, split when the male of a pair in the next burrow died and the female sought a new mate.

Buddy and Pedro, a pair of male African Penguins, will be separated by the Toronto Zoo to mate with female penguins.

Vultures: In 1998 two male Griffon vultures named Dashik and Yehuda, at the Jerusalem Biblical Zoo, engaged in "open and energetic sex" and built a nest. The keepers provided the couple with an artificial egg, which the two parents took turns incubating; and 45 days later, the zoo replaced the egg with a baby vulture. The two male vultures raised the chick together. A few years later, however, Yehuda became interested in a female vulture that was brought into the aviary. Dashik became depressed, and was eventually moved to the zoological research garden at Tel Aviv University where he too set up a nest with a female vulture.

Two homosexual male vultures at the Allwetter Zoo in Muenster built a nest together, although they were picked on and often had their nest materials stolen by other vultures. They were eventually separated to try to promote breeding by placing one of them with female vultures, despite the protests of German homosexual groups.

Pigeons: Both male and female pigeons sometimes exhibit homosexual behaviour. As well as sexual behaviour same-sex pigeon pairs will build nests, and hens will lay (infertile) eggs and attempt to incubate them. Some pigeons also display fetish behaviour and attempt to mate with specific inanimate objects.

Mammals

Amazon Dolphin: The Amazon River dolphin or boto has been reported to form up in bands of 3–5 individuals enjoying group sex. The groups usually comprise young males and sometimes one or two females. Sex is often performed in non-reproductive ways, using snout, flippers and genital rubbing, without regards to gender. In captivity, they have been observed to sometimes perform homosexual and heterosexual penetration of the blowhole, a hole homologous with the nostril of other mammals, making this the only known example of nasal sex in the animal kingdom. The males will sometimes also perform sex with tucuxi males, a small porpoise.

American Bison: Courtship, mounting, and full anal penetration between bulls has been noted to occur among American Bison. The Mandan nation Okipa festival concludes with a ceremonial enactment of this behaviour, to "ensure the return of the buffalo in the coming season." Also, mounting of one female by another is common among cattle.

Bonobo and Other Apes: The Bonobo, which has a matriarchal society, unusual amongst apes, is a fully bisexual species—both males and females engage in heterosexual and homosexual behaviour, being noted for female-female homosexuality in particular. About 60% of all sexual activity in this species is between two or more females. While the homosexual bonding system in Bonobos represents the highest frequency of homosexuality known in any species, homosexuality has been reported for all great apes (a group which includes humans), as well as a number of other primate species. Dutch primatologist Frans de Waal on observing and filming bonobos noted that there were two reasons to believe sexual activity is the bonobo's answer to avoiding conflict.

Anything that arouses the interest of more than one bonobo at a time, not just food, tends to result in sexual contact. If two bonobos approach a cardboard box thrown into their enclosure, they will briefly mount each other before playing with the box. Such situations lead to squabbles in most other species. But bonobos are quite tolerant, perhaps because they use sex to divert attention and to defuse tension.

Bonobo sex often occurs in aggressive contexts totally unrelated to food. A jealous male might chase another away from a female, after which the two males reunite and engage in scrotal rubbing. Or after a female hits a juvenile, the latter's mother may lunge at the aggressor, an action that is immediately followed by genital rubbing between the two adults.

Bottlenose Dolphins: Dolphins of several species engage in homosexual acts, though it is best studied in the bottlenose dolphins. Sexual encounters between females take the shape of "beak-genital propulsion", where one female insert her beak in the genital opening of the other while swimming gently forward. Between males, homosexual behaviour include rubbing of genitals against each other, which sometimes lead to the males swimming belly to belly, inserting the penis in the others genital slit and sometimes anus.

Janet Mann, Georgetown University professor of biology and psychology, argues that the strong personal behaviour among male dolphin calves is about bond formation and benefits the species in an evolutionary context. She cites studies showing that these dolphins later in life as adults are in a sense bisexual, and the male bonds forged earlier in life work together for protection as well as locating females to reproduce with. Confrontations between flocks of bottlenose dolphins and the related species Atlantic spotted dolphin will sometimes lead to cross-species homosexual behaviour between the males rather than combat.

Elephants: African and Asian males will engage in same-sex bonding and mounting. Such encounters are often associated with affectionate interactions, such as kissing, trunk intertwining, and placing trunks in each other's mouths. Male elephants, who often live apart from the general herd, often form "companionships", consisting of an older individual and one or sometimes two younger, attendant males with sexual behaviour being an important part of the social dynamic. Unlike heterosexual relations, which are always of a fleeting nature, the relationships between males may last for years. The encounters are analogous to heterosexual bouts, one male often extending his trunk along the other's back and pushing forward with his tusks to signify his intention to mount. Same-sex relations are common and frequent in both sexes, with Asiatic elephants in captivity devoting roughly 45% of sexual encounters to same-sex activity.

Giraffes

Male giraffes have been observed to engage in remarkably high frequencies of homosexual behaviour. After aggressive "necking", it is common for two male giraffes to caress and court each other, leading up to mounting and climax. Such interactions between males have been found to be more frequent than heterosexual coupling. In one study, up to 94% of observed mounting incidents took place between two males. The proportion of same sex activities varied between 30 and 75%, and at any given time one in twenty males were engaged in non-combative necking behaviour with another male. Only 1% of same-sex mounting incidents occurred between females.

Humans

Monkeys: Among monkeys, Lionel Tiger and Robin Fox conducted a study on how Depo-Provera contraceptives lead to decreased male attractiveness to females and eventually to male homosexuality. Janet E. Smith summarizes the findings as follows:

The study in the early 70s, involved a tribe of monkeys. The alpha monkey of this tribe, named Austin, chose three female monkeys to be his exclusive sexual partners. Austin had a grand time with these three female monkeys. Then the researchers injected Austin's three females with the contraceptive Depo-Provera. Austin stopped having sex with them and chose other female monkeys to be his sexual partners. Then they contracepted all of the females in the tribe. The males stopped have sex with the females and started behaving in a turbulent and confused manner.

Japanese Macaque: With the Japanese macaque, also known as the "snow monkey", same-sex relations are frequent, though rates vary between troops. Females will form "consortships" characterised by affectionate social and sexual activities. In some troops up to one quarter of the females form such bonds, which vary in duration from a few days to a few weeks. Often, strong and lasting friendships result from such pairings. Males also have same-sex relations, typically with multiple partners of the same age. Affectionate and playful activities are associated with such relations.

Lions: Both male and female lions have been seen to interact homosexually. Male lions pair-bond for a number of days and initiate homosexual activity with affectionate nuzzling and caressing, leading to mounting and thrusting. About 8% of mountings have been observed to occur with other males. Pairings between females are held to be fairly common in captivity but have not been observed in the wild.

Polecat: European polecats *Mustela putorius* were found to engage homosexually with non-sibling animals. Exclusive homosexuality with mounting and anal penetration in this solitary species serves no apparent adaptive function.

Sheep: The reason why *Ovis aries* has attracted so much attention is that some rams seem to have an exclusive homosexual orientation.

An October 2003 study by Dr. Charles E. Roselli et al. (Oregon Health and Science University) states that homosexuality in male sheep (found in 8% of rams) is associated with a region in the rams' brains which the authors call the "ovine Sexually Dimorphic Nucleus" (oSDN) which is half the size of the corresponding region in heterosexual male sheep.

Scientists found that, "The oSDN in rams that preferred females was significantly larger and contained more neurons than in male-oriented rams and ewes. In addition, the oSDN of the female-oriented rams expressed higher levels of aromatase, a substance that converts testosterone to estradiol, a form of estrogen which is believed to facilitate typical male sexual behaviours. Aromatase expression was no different between male-oriented rams and ewes."

"The dense cluster of neurons that comprise the oSDN express cytochrome P450 aromatase. Aromatase mRNA levels in the oSDN were significantly greater in female-oriented rams than in ewes, whereas male-oriented rams exhibited intermediate levels of expression." These results suggest that "...naturally occurring variations in sexual partner preferences may be related to differences in brain anatomy and its capacity for estrogen synthesis." As noted prior, given the potential unagressiveness of the male population in question, the differing aromatase levels may also have been evidence of aggression levels, not sexuality. It should also be noted that the results of this study have not been confirmed by other studies.

The Merck Manual of Veterinary Medicine appears to consider homosexuality among sheep as a routine occurrence and an issue to be dealt with as a problem of animal husbandry.

Spotted Hyena: The family structure of the Spotted Hyena is matriarchal, and dominance relationships with strong sexual elements are routinely observed between related females. Due largely to the female spotted hyena's unique urogenital system, which looks more like a penis rather than a vagina, early naturalists thought hyenas were hermaphroditic males who commonly practiced homosexuality. Early writings such as Ovid's *Metamorphoses* and the *Physiologus* suggested that the hyena continually changed its sex and nature from male to female and back again. In *Paedagogus*, Clement of Alexandria noted that the hyena (along with the hare) was "quite obsessed with sexual intercourse." Many Europeans associated the hyena with sexual deformity, prostitution, deviant sexual behaviour, and even witchcraft.

The reality behind the confusing reports is the sexually aggressive behaviour between the females, including mounting between females. Research has shown that "in contrast to most other female mammals, female *Crocuta* are male-like in appearance, larger than males, and substantially more aggressive," and they have "been masculinised without being defeminised."

Study of this unique genitalia and aggressive behaviour in the female hyena has led to the understanding that more aggressive females are better able to compete for resources, including food and mating partners. Research has shown that "elevated levels of testosterone in utero" contribute to extra aggressiveness; both males and females mount members of the same sex, who in turn are possibly acting more submissive because of lower levels of testosterone in utero.

Others

Lizards: Whiptail lizard (Teiidae genus) females have the ability to reproduce through parthenogenesis and as such males are rare and sexual breeding non-standard. Females engage in sexual behaviour to stimulate ovulation, with their behaviour following their hormonal cycles; during low levels of estrogen, these (female) lizards engage in "masculine" sexual roles. Those animals with currently high estrogen levels assume "feminine" sexual roles.

Lizards that perform the courtship ritual have greater fertility than those kept in isolation due to an increase in hormones triggered by the sexual behaviours. So, even though asexual whiptail lizards populations lack males, sexual stimuli still increase reproductive success.

From an evolutionary standpoint, these females are passing their full genetic code to all of their offspring (rather than the 50% of genes that would be passed in sexual reproduction). Certain species of gecko also reproduce by parthenogenesis.

Dragonflies

Male homosexuality has been inferred in several species of dragonflies (the order Odonata). The cloacal pinchers of male damselflies and dragonflies inflict characteristic head damage to females during sex. A survey of 11 species of damsel and dragonflies has revealed such mating damages in 20 to 80 % of the males too, indicating a fairly high occurrence of sexual coupling between males.

Fruit Flies

Male *Drosophila melanogaster* flies bearing two copies of a mutant allele in the fruitless gene court and attempt to mate exclusively with other males. The genetic basis of animal homosexuality has been studied in the fly *Drosophila melanogaster*. Here, multiple genes have been identified that can cause homosexual courtship and mating. These genes are thought to control behaviour through pheromones as well as altering the structure of the animal's brains. These studies have also investigated the influence of environment on the likelihood of flies displaying homosexual behaviour.

Bed Bugs

Male bed bugs (Cimex lectularius) are sexually attracted to any newly fed individual and this results in homosexual mounting. This occurs in heterosexual mounting by the traumatic insemination in which the male pierces the female abdomen with his needle-like penis. In homosexual mating this risks abdominal injuries as males lack the female counteradaptive spermalege structure. Males produce alarm pheromones to reduce such homosexual matings.

Sequential Hermaphroditism

Sequential hermaphroditism (called dichogamy in botany) is a type of hermaphroditism that occurs in many fish, gastropods and plants. Here, the individual is born one sex and changes sex at some point in their life. They can change from a male to female (protandry), or from female to male (protogyny). Despite which sex the organism changes to, those that change gonodal sex can have both female and male germ cells in the gonads or can change from one complete gonodal type to the other during their last life stage.

Zoology

Protandry: Protandry refers to organisms that are born male and at some point in their lifespan change sex to female. Protandrous animals include clownfish. Clownfish have a very structured society. In the *Amphiprion percula* species, there are zero to four individuals excluded from breeding and a breeding pair living in a sea anemone. Dominance is based on size, the female being the largest and the male being the second largest.

The rest of the group is made up of progressively smaller non-breeders, which have no functioning gonads. If the female dies, the male gains weight and becomes the female for that group. The largest non-breeding fish then sexually matures and becomes the male of the group.

Other examples of protandrous animals include:

- The ctenophore *Coeloplana gonoctena.* In this organism the females are bigger than the males and are only found during the summer. In contrast males are found year round.
- The flatworms *Hymanella retenuova* and *Paravortex cardii.*
- *Laevapex fuscus*, a gastropod, is described as being functionally protandric. The sperm matures in late winter and early spring, and the eggs mature in early summer, and copulation occurs only in June. This shows that males cannot reproduce until the females appear, thus why they are considered to be functionally protandric.

Protogyny

Protogyny refers to organisms that are born female and at some point in their lifespan change sex to males. Common model organisms for this type of sequential hermaphroditism are wrasses. They are one of the largest families of coral reef fish and belong to the Labridae family. Wrasses are found around the world in all marine habitats and tend to bury themselves in sand at night or when they feel threatened. In wrasses, the larger of the two fish is the male, while the smaller is the female. In most cases, females and immature have a uniform colour while the male has the terminal bicoloured phase. Large males hold territories and try to pair spawn while small to mid-size initial-phase males live with females and group spawn. In other words, both the initial and terminal phase males can breed; they differ however in the way they do it.

In the California Sheephead (*Pimelometopon pulchrum*), a type of wrasse, when the female changes to male, the ovaries degenerate and spermatogenic crypts appear in the gonads. The general structure of the gonads remains ovarian after the transformation and the sperm is transported through a series of ducts on the periphery of the gonad and oviduct. Here sex change is age dependant. For example, the California sheephead stays a female for four years before changing sex.

Other examples of protogynous organisms include:

- The isopods *Cyathura polita* and *C. carinata*
- The tanaidacean *Heterotanais oerstedi.*
- The echinoderms, *Asterina pancerii* and *A. gibbosa* are also protogynous and they brood their young.
- Protogyny sometimes occurs in the frog *Rana temporaria*, where old females sometimes change to males.

Ultimate Causes

Ghiselin proposed three models for hermaphroditism in 1969 in his paper titled "The evolution of hermaphroditism among animals". The "low-density model" states that individuals have characteristics that reduce the opportunity for mating; this model cannot be applied to sequential hermaphroditism.

The "gene dispersal model" is based on the idea that limitations on dispersal may influence population structure or genetical environment and it can be separated into two versions: the inbreeding version and the sampling-error version. This theory of gene dispersal can be applied to sequential hermaphrodites, especially the inbreeding version. The inbreeding version is based upon the fact that both protandry and protogyny help prevent inbreeding in plants and thus one can make the same assumption that in animals it works by reducing the probability of this occurring among siblings.

The sampling-error version is based on the reality that the genetical environment is influenced by genetic drift and similar phenomena in small populations. The two aspects of these hypotheses influenced by hermaphroditism, that is inbreeding and sampling-error, result in the same thing, reduction of genetic variability. In other words being a hermaphrodite would increase genetic variability and thus be considered advantageous to the organism. This theory of gene dispersal can be applied to sequential hermaphrodites, especially the inbreeding version. Lastly, the "size-advantage model" states that reproductive functions are carried out better if the individual is a certain size/age. Assuming that the reproductive functions of one sex are better performed at a certain size, then an organism would assume the sex that its size allows to perform the best. This would increase its reproductive potential and fitness. For example, eggs are larger than sperm, thus if you are a big you are able to make more eggs so being female when big is advantageous, however the size advantage relationship is really not as simple as the example just mentioned, but it allows for a better understanding of it.

In most ectotherms body size and female fecundity are positively correlated. This supports Ghiselin's size-advantage model, which is still widely accepted today. Kazancioglu and Alonzo (2010) performed the first comparative analysis of sex change in Labridae. Their analysis supports the size-advantage model by Ghiselin and suggest that sequential hermaphroditism is correlated to the size-advantage. They determined that dioecy was less likely to occur when the size advantage is stronger than other advantages

Warner suggests that selection for protandry may occur in populations where female fecundity is augmented with age and individuals mate randomly. Selection for protogyny may occur where there are traits in the population that depress male fecundity at early ages (territoriality, mate selection or inexperience) and when female fecundity is decreased with age, the latter seems to be rare in the field. An example of territoriality favouring protogyny occurs when there is a need to protect their habitat and being a large male is advantageous for this purpose. In the mating aspect, a large male has a higher chance of mating, while this has no effect on the female mating fitness. Thus, he suggests that female fecundity has more impact on sequential hermaphroditism that the age structures of the population.

The size-advantage model predicts that sex change would only be absent if the relationship between size/age with reproductive potential is identical in both sexes.

With this prediction one would assume that hermaphroditism is very common, but this is not the case. Sequential hermaphroditism is very rare and according to scientists this is due to some cost that decreases fitness in sex changers as opposed to those who don't change sex. Kazanciglu and Alonzo confirmed this in 2009. They found that the costs of changing sex only favoured dioecy when the cost was very large but that some groups favoured hermaphroditism. This indicates that the cost of sex change does not explain the rarity of sequential hermaphroditism by itself.

Proximate Causes

Many studies have focused on the proximate causes of sequential hermaphroditism. The role of aromatase has been widely studied in this area. Aromatase is an enzyme that controls the androgen/estrogen ratio in animals by catalysing the conversion of testosterone into oestradiol, which is irreversible. It has been discovered that the aromatase pathway mediates sex change in both directions. Many studies also involve understanding the effect of aromatase inhibitors on sex change. One such study was performed by Kobayashi et al. In their study they tested the role of estrogens in male three-spot wrasses (*Halichoeres trimaculatus*). They discovered that fish treated with aromatase inhibitors showed decreased gonodal weight, plasma estrogen level and spermatogonial proliferation in the testis as well as increased androgen levels. Their results suggest that estrogens are important in the regulation of spermatogenesis in this protogynous hermaphrodite.

Animals Displaying Homosexual Behaviour

Animals (birds, mammals, insects, reptiles, fish etc.) for which there is documented evidence of homosexual or transgender behaviour of one or more of the following kinds: sex, courtship, affection, pair bonding, or parenting, as noted in researcher and author Bruce Bagemihl's 1999 book *Biological Exuberance: Animal Homosexuality and Natural Diversity*.Bagemihl writes that the presence of same-sex sexual behaviour was not 'officially' observed on a large scale until the 1990s due to possible observer bias caused by social attitudes towards LGBT people making the homosexual theme taboo. Bagemihl devotes three chapters; *Two Hundred Years at Looking at Homosexual Wildlife, Explaining (Away) Animal Homosexuality* and *Not For Breeding Only* in his 1999 book *Biological Exuberance* to the "documentation of systematic prejudices" where he notes "the *present ignorance* of biology lies precisely in its single-minded attempt to find reproductive (or other) "explanations" for homosexuality, transgender, and non-procreative and alternative heterosexualities.

Petter Bøckman, academic adviser for the *Against Nature?* Exhibit stated "Many researchers have described homosexuality as something altogether different from sex. They must realise that animals can have sex with who they will, when they will and without consideration to a researcher's ethical principles". Homosexual behaviour is found amongst social birds and mammals, particularly the sea mammals and the primates. Animal sexual behaviour takes many different forms, even within the same species and the motivations for and implications of their behaviours have yet to be fully understood. Bagemihl's research shows that homosexual behaviour, not necessarily sex, has been observed in close to 1500 species, ranging from primates to gut worms, and is well documented for 500 of them.

Homosexuality in animals is seen as controversial by social conservatives because it asserts the naturalness of homosexuality in humans, while others counter that it has no implications and is nonsensical to equate animal behaviour to morality. Animal preference and motivation is always inferred from behaviour. Thus homosexual behaviour has been given a number of terms over the years. The correct usage of the term *homosexual* is that an animal *exhibits homosexual behaviour*, however this article conforms to the usage by modern research applying the term *homosexuality* to all sexual behaviour (copulation, genital stimulation, mating games and sexual display behaviour) between animals of the same sex.

The all-female Whiptail lizard species *Cnemidophorus neomexicanus* (centre), which reproduces via parthenogenesis, is shown flanked by two sexual species having males, *C. inornatus* (left) and *C. tigris* (right). Research has shown that simulated mating behaviour increases fertility for *Cnemidophorus neomexicanus*. One female lies on top of another, playing the role of the male, the lizard that was on bottom has larger eggs. The lizards switch off this role each mating season.

Male homosexuality has been inferred in several species of dragonflies. A survey of damsel and dragonflies reveals characteristic cloacal pincher mating damage in 20–80 percent of the males, indicating a fairly high occurrence of sexual coupling between males.

Mammals

Selected mammals from the full list:

- Bison
- Brown Bear
- Brown Rat

- Caribou
- Cat (domestic)
- Cattle (domestic)
- Cheetah
- Chimpanzee
- Common Dolphin
- Common Marmoset
- Dog
- Elephant
- Fox
- Giraffe
  Goat
- Horse (domestic)
- Human
- Koala
- Lion
- Orca
- Raccoon.

Birds

Selected birds from the full list:

- Barn Owl
- Chicken
- Common Gull
- Emu

- King Penguin
- Mallard
- Raven
- Seagull.

Fish

- Amazon molly
- Blackstripe topminnow
- Bluegill Sunfish
- Char

- Grayling
- European Bitterling
- Green swordtail
- Guiana leaffish
- Houting Whitefish
- Jewel Fish
- Least Darter (*Microperca punctulata*)
- Mouthbreeding Fish sp.
- Salmon spp.
- Southern platyfish
- Ten-spined stickleback
- Three-spined stickleback.

Reptiles

- Anole sp.
- Bearded Dragon
- Broad-headed Skink
- Checkered Whiptail Lizard
- Chihuahuan Spotted Whiptail Lizard
- Common Ameiva
- Common Garter Snake
- Cuban Green Anole
- Desert Grassland Whiptail Lizard
- Desert Tortoise
- Fence Lizard
- Five-lined Skink
- Gopher (Pine) Snake
- Green Anole
- Inagua Curlytail Lizard
- Jamaican Giant Anole
- Laredo Striped Whiptail Lizard
- Largehead Anole
- Mourning Gecko
- Plateau Striped Whiptail Lizard
- Red Diamond Rattlesnake

- Red-tailed Skink
- Side-blotched Lizard
- Speckled Rattlesnake
- Water Moccasin
- Western rattlesnake (Crotalus viridis)
- Western Banded Gecko
- Whiptail Lizard spp.
- Wood Turtle.

Amphibians

- Appalachian Woodland Salamander
- Black-spotted Frog
- Mountain Dusky Salamander
- Tengger Desert Toad.

Insects

- Alfalfa Weevil
- Australian Parasitic Wasp sp.
- Bean Weevil sp.
- Bedbug and other Bug spp.
- Blister Beetle spp.
- Blowfly
- Broadwinged Damselfly sp.
- Cabbage (Small) White (Butterfly)
- Checkerspot Butterfly
- Club-tailed Dragonfly spp.
- Cockroach spp.
- Common Skimmer Dragonfly spp.
- Creeping Water Bug sp.
- Cutworm
- Digger Bee
- Dragonfly spp.
- Eastern Giant Ichneumon Wasp
- Eucalyptus Longhorned Borer
- Field Cricket sp.
- Flour Beetle

- Fruit Fly spp.
- Glasswing Butterfly
- Grape Berry Moth
- Grape Borer
- Green Lacewing
- Hen Flea
- House Fly
- Ichneumon wasp sp.
- Japanese Scarab Beetle
- Larch Bud Moth
- Large Milkweed Bug
- Large White
- Long-legged Fly spp.
- Mazarine Blue
- Mediterranean Fruit Fly
- Mexican White (butterfly)
- Midge sp.
- Migratory Locust
- Monarch Butterfly
- Narrow-winged Damselfly spp.
- Parsnip Leaf Miner
- Pomace fly
- Queen Butterfly
- Red Ant sp.
- Red Flour Beetle
- Reindeer Warble Fly (Hypoderma tarandi)
- Rose Chafer
- Rove Beetle spp.
- Scarab Beetle (Melolonthine)
- Screwworm Fly
- Silkworm Moth
- Southeastern Blueberry Bee
- Southern Green Stink Bug
- Southern Masked Chafer

- Southern One-Year Canegrub
- Spreadwinged Damselfly spp.
- Spruce Budworm Moth
- Stable Fly sp.
- Stag Beetle spp.
- Tsetse Fly
- Water Boatman Bug
- Water Strider spp.

Other Invertebrates

- Blood-Fluke
- Box Crab
- Harvest Spider sp.
- Hawaiian Orb-Weaver (spider)
- Incirrate Octopus spp.
- Jumping Spider sp.
- Mite sp.
- Spiny-Headed Worm.

Chapter 9

Sexual Dimorphism

Sexual dimorphism is a phenotypic difference between males and females of the same species. Examples of such differences include differences in morphology, ornamentation, and behaviour.

***Figure:** Female (left) and male Common Pheasant, illustrating the dramatic difference in both colour and size between sexes*

Ornamentation/colouration

A common type of dimorphism is ornamentation. Dichromatism is where a sex of a given species is more than one colour including that of saturated, vibrant hues.

Exaggerated dimorphic traits are used predominantly in the competition over mates. Ornaments are costly to produce or maintain. but differ depending on the type of colour mechanism involved.

One excellent example of this is in the peacock. The peacock has a very ornate plumage which it uses to attract peahens. The peacock will use its ornate plumage to court a mate. At first look, one may mistake a peacock as a completely different species from the peahen because of its vibrant colours and the sheer size of the male's plumage.

Figure: *The peacock, on the right, is courting the peahen, on the left.*

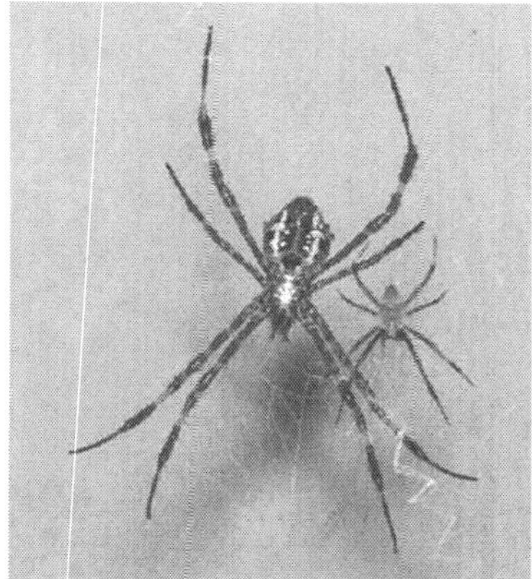

Figure: *Female (left) and male Argiope appensa, displaying typical sexual dimorphism differences in spiders where the male is dramatically smaller.*

The peahen is more of a dullish brown colour and lacks the ornate plumage. Although the plumage of the peacock is helpful when courting a mate, it is actually detrimental to the survival of the peacock. The large plumage makes it nearly impossible for the peacock to fly. The peacock is limited to short bursts of flight when they attempt to escape predators by taking refuge in trees. The feathers' bright colors make it harder for the bird to remain unseen, not only by predators but also by humans, who see the feathers as a desired item.

Another example of sexual dichromatism is that of the nestling blue tits. Males are chromatically more yellow than females. It is believed that this is obtained by the ingestion of green lepidopteran larvae, which contain large amounts of the carotenoids lutein and zeaxanthin. This diet also affects the sexually dimorphic colors in the human-invisible UV spectrum. In other words, the male birds, although appearing yellow to humans actually have a violet-tinted plumage that is seen by females. This plumage is thought to be an indicator of male parental abilities. Perhaps this is a good indicator for females because it shows that they are good at obtaining a food supply which the carotenoid is obtained from.

There is a positive correlation between the chromas of the tail and breast feathers and body condition. Carotenoids play an important role in portraying immune function for many animals. More carotenoids mean a higher immune resistance. Those with more yellow and blue pigmentation were also found to be heavier.

In many instances, females show preference for exaggerated male secondary sexual characteristics when choosing a mate (Ryan and Rand, 1993). Females tend to show direction preferences for more elaborate males. Females have been shown to discriminate against males which are dull in colour. There have also been species such as estrildid finch where premating isolation was seen due to lack of vibrant colours by the males. This female preference for ornamentation may affect the evolution of discriminatory mating preferences. This is known as the ornamentation hypothesis.

Species with Larger Females than Males

Figure: *Orgyia recens above male, below female*

In some species such as insects, spiders, many fish, birds of prey and certain mammals such as the spotted hyena and the blue whale, the female is larger than the male. As an example, in some species females are sedentary and sparsely distributed, and so males must search for them. Vollrath and Parker argue that this difference in behaviour leads to radically different selection pressures on the two sexes, evidently favouring smaller males. Cases where the male is larger than the female have been studied as well, and require alternate explanations.

One example of sexual size dimorphism is the bat *Myotis nigricans.* In this species, females are substantially larger than males. They differ in body weight, skull measurement, and forearm length. The difference in size is believed to be caused by natural selection for a large female size due to a fecundity advantage. The interaction between the sexes and energetic needs such as time and energy required to produce viable offspring make it favorable for females to be larger in this species. Females bear the energetic cost of producing eggs which is much greater than that of the male who only bears the cost of making sperm. The fecundity advantage hypothesis states, that a big mother is able to produce more offspring and give those offspring more favorable conditions to ensure their survival. This is true for most ectotherms. Another reason why females are believed to be larger is due to the fact that they provide parental care for a substantial amount of time while the offspring matures. The time of gestation and lactation is fairly long in the *M. nigricans*, where females suckle their offspring until nearly adult size. They would not be able to fly and catch prey if they did not compensate for the additional mass of the offspring during this time.In addition to the hypothesis that explains an advantage of large female size, it is hypothesized that smaller male size is an adaptation for males to increase maneuverability and agility. This selection for agility in flying is a helpful adaptation which allows males to better compete with females for food and other resources.

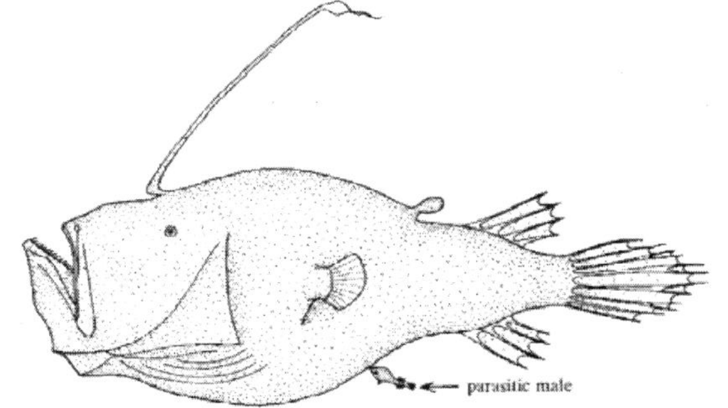

Figure: *Female triplewart seadevil, an anglerfish, with male attached near vent (arrow)*

Some species of anglerfish also display extreme sexual dimorphism. Females are more typical in appearance to other fish, whereas the males are tiny rudimentary creatures with stunted digestive systems. A male must find a female and fuse with her: he then lives parasitically, becoming little more than a sperm-producing body. A similar situation is found in the Zeus water bug *Phoreticovelia disparata* where the female has a glandular area on her back that can serve to feed a male that clings to her (note that although males can survive away from females, they generally are not free-living).

Psychological and Behavioural Differentiation

Sex steroid-induced differentiation of adult reproductive and other behaviour has been demonstrated experimentally in many animals. In some mammals, adult sex-dimorphic reproductive behaviour (e.g., mounting or receptive lordosis) can be shifted to that of the other sex by supplementation or deprivation of androgens in fetal life or early infancy, even if adult levels are normal.

Evolution of Sexual Dimorphism

In 1871 Charles Darwin advanced the theory of sexual selection, which related sexual dimorphism with sexual selection.

In many non-monogamous species, the benefit to a male's reproductive fitness of mating with multiple females is large, whereas the benefit to a female's reproductive fitness of mating with multiple males is small or non-existent. In these species, there is a selection pressure for whatever traits enable a male to have more matings. The male may therefore come to have different traits from the female.

These traits could be ones that allow him to fight off other males for control of territory or a harem, such as large size or weapons; or they could be traits that females, for whatever reason, prefer in mates. Male-male competition poses no deep theoretical questions but female choice does. Females may choose males that appear strong and healthy, thus likely to possess "good alleles" and give rise to healthy offspring. However, in some species females seem to choose males with traits that do not improve offspring survival rates, and even traits that reduce it (potentially leading to traits like the peacock's tail). Two hypotheses for explaining this fact are the sexy son hypothesis and the handicap principle.

The sexy son hypothesis states that females may initially choose a trait because it improves the survival of their young, but once this preference has become widespread, females must continue to choose the trait, even if it becomes harmful. Those that do not will have sons that are unattractive to most females (since the preference is widespread) and so receive few matings.

The handicap principle states that a male who survives despite possessing some sort of handicap thus proves that the rest of his genes are "good alleles." If males with "bad alleles" could not survive the handicap, females may evolve to choose males with this sort of handicap; the trait is acting as a hard-to-fake signal of fitness.

Sexual Dimorphism in Mammals and Birds

The brains of many birds and mammals, including humans, are significantly different for males and females of the species. Both genes and hormones affect the formation of many animal brains before "birth" (or hatching), and also behaviour of adult individuals. Hormones significantly affect human brain formation, and also brain development at puberty. A 2004 review in *Nature Reviews Neuroscience* observed that "because it is easier to manipulate hormone levels than the expression of sex chromosome genes, the effects of hormones have been studied much more extensively, and are much better understood, than the direct actions in the brain of sex chromosome genes." It concluded that while "the differentiating effects of gonadal secretions seem to be dominant," the existing body of research "support the idea that sex differences in neural expression of X and Y genes significantly contribute to sex differences in brain functions and disease."

Birds

Sexual dimorphism in birds can be manifested in size or plumage differences between the sexes. Sexual size dimorphism varies among taxa with males typically being larger, though this is not always the case i.e. birds of prey and some species of flightless birds. Plumage dimorphism, in the form of ornamentation or colouration, also varies, though males are typically the more ornamented or brightly coloured sex. Such differences have been attributed to the unequal reproductive contributions of the sexes. In some species, the male's contribution to reproduction ends at copulation, while in other species the male becomes the main caregiver. Plumage polymorphisms have evolved to reflect these differences and other measures of reproductive fitness, such as body condition or survival. The male phenotype sends signals to females who then choose the 'fittest' available male.

Sexual dimorphism is a product of both genetics and environmental factors. An example of sexual polymorphism determined by environmental conditions exists in the house finch. House finch males can be classified into three categories during breeding season: black breeders, brown breeders, and brown auxiliaries. These differences arise in response to the bird's body condition: if they are healthy they will produce more androgens thus becoming black breeders, while less healthy birds produce less androgens and become brown auxiliaries. The reproductive success of the male is thus determined by his success during each year's non-breeding season, causing reproductive success to vary with each year's environmental conditions.

Sexual dimorphism is maintained by the counteracting pressures of natural selection and sexual selection. For example, sexual dimorphism in colouration increases the vulnerability of bird species to predation by European sparrowhawks in Denmark. Presumably, increased sexual dimorphism means males are brighter and more conspicuous, leading to increased predation. Moreover, the production of more exaggerated ornaments in males may come at the cost of suppressed immune function. So long as the reproductive benefits of the trait due to sexual selection are greater than the costs imposed by natural selection, then the trait will propagate throughout the population. Reproductive benefits arise in the form of a larger number of offspring, while natural selection imposes costs in the form of reduced survival. This means that even if the trait causes males to die earlier, the trait is still beneficial so long as males with the trait produce more offspring than males lacking the trait.

Such differences in form and reproductive roles often cause differences in behaviour. As previously stated, males and females often have different roles in reproduction. The courtship and mating behaviour of males and females are regulated largely by hormones throughout a bird's lifetime. Activational hormones occur during puberty and adulthood and serve to 'activate' certain behaviours when appropriate, such as territoriality during breeding season. Organisational hormones occur only during a critical period early in development, either just before or just after hatching in most birds, and determine patterns of behaviour for the rest of the bird's life. Such behavioural differences can cause disproportionate sensitivities to anthropogenic pressures. Females of the whinchat in Switzerland breed in intensely managed grasslands. Earlier harvesting of the grasses during the breeding season lead to more female deaths. Populations of many birds are often male-skewed and when sexual differences in behaviour increase this ratio, populations decline at a more rapid rate.

Consequently, sexual dimorphism has important ramifications for conservation. However, sexual dimorphism is not only found in birds and is thus important to the conservation of many animals. Such differences in form and behaviour can lead to sexual segregation, defined as sex differences in space and resource use. Most sexual segregation research has been done on ungulates, but such research extends to bats, kangaroos, and birds. Sex-specific conservation plans have even been suggested for species with pronounced sexual segregation.

Fish

There are also cases where males are substantially larger than that of females. An example is *Lamprologus callipterus*, a type of cichlid fish. In this fish, the males are characterised as being up to 60 times larger than that of the females. The males increased size is believed to be advantageous because males collect and defend empty snail shells in each of which a female breeds. Males must be larger and more powerful in order to collect the largest shells. Female's body size must remain small because in order for her to breed, she must lay her eggs inside of the empty shells. If she grows too large, she will not fit in the shells and be unable to breed.

The female's small body size is likely beneficial in her success if finding an unoccupied shell. Larger shells, although preferred by females, are often limited in availability. Hence, the female is limited to the growth of the size of the shell and may actually change her growth rate according to shell size availability. In other words, the male's ability to collect large shells depends on his size. The larger the male, the larger the shells he is able to collect. This then allows for females to be larger in his brooding nest which makes the difference between the sizes of the sexes less substantial. Male-male competition in this fish species also selects for large size in males. There is aggressive competition by males over territory and access to larger shells. Large males win fights and steal shells from competitors. Sexual dimorphism also occurs in hermaphroditic fish. These species are known as sequential hermaphrodites. In fish, reproductive histories often include the sex-change from female to male where there is a strong connection between growth, the sex of an individual, and the mating system it operates within. In protogynous mating systems where males dominate mating with many females, size plays a significant role in male reproductive success. Males have a propensity to be larger than females of a comparable age but it is unclear whether the size increase is due to a growth spurt at the time of the sexual transition or due to the history of faster growth in sex changing individuals. Larger males are able to stifle the growth of females and control environmental resources.

Social organisation plays a large role in the changing of sex by the fish. It is often seen that a fish will change its sex when there is a lack of dominant male within the social hierarchy. The females that change sex are often those who attain and preserve an initial size advantage early in life. In either case, females which change sex to males are larger and prove to be a good example of dimorphism.

Balancing Selection

Balancing selection refers to a number of selective processes by which multiple alleles (different versions of a gene) are actively maintained in the gene pool of a population *at frequencies above that of gene mutation*. This usually happens when the heterozygotes for the alleles under consideration have a higher adaptive value than the homozygote. In this way genetic polymorphism is conserved.

Figure: *Biston betularia morpha typica, the standard light-coloured Peppered Moth.*

Figure: *Biston betularia morpha carbonaria, the melanic Peppered Moth. The proportions of this form vary in different locations*

There are three main types of natural selection: In directional selection the allele frequency for a trait continuously shifts in one direction. In stabilising selection the frequency of the alleles of lower fitness decreases until they vanish. Balancing selection is similar but not identical to disruptive selection where individuals of extreme trait values are favored against those with average trait values. These terms are used for quantitative characters controlled by a number of genes.

Evidence for balancing selection can be found in the number of alleles in a population which are maintained above mutation rate frequencies. All modern research has shown that this significant genetic variation is ubiquitous in panmictic populations. It is a genetic expression of the field experience of Darwin, Wallace and others, that natural populations in the wild are extraordinarily varied (though not all such variation is of genetic origin).

There are several mechanisms (which are not exclusive within any given population) by which balancing selection works to maintain polymorphism. The two major and most studied are *heterozygote advantage* and *frequency-dependent selection.*

Mechanisms of Balancing Selection

Heterozygote Advantage: In heterozygote advantage, or heterotic balancing selection, an individual who is heterozygous at a particular gene locus has a greater fitness than a homozygous individual. Polymorphisms maintained by this mechanism are balanced polymorphisms.

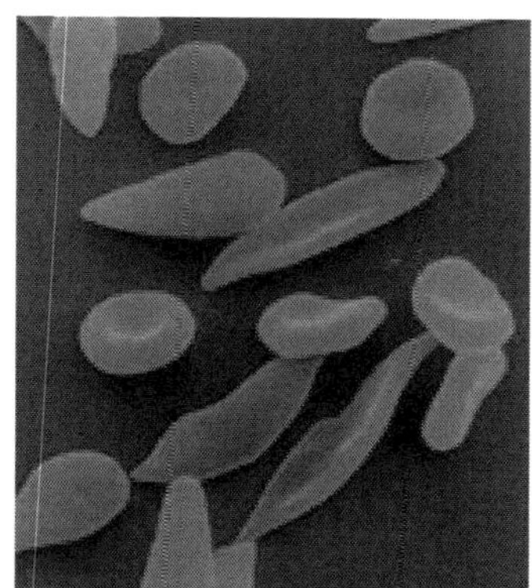

Figure: *Sickle-shaped red blood cells. This non-lethal condition in heterozygotes is maintained by balancing selection in humans of Africa and India due to its resistance to the malarial parasite.*

A well-studied case is that of sickle cell anemia in humans, a hereditary disease that damages red blood cells. Sickle cell anemia is caused by the inheritance of a variant hemoglobin gene (HgbS) from both parents. In these individuals, hemoglobin in red blood cells is extremely sensitive to oxygen deprivation, and this causes shorter life expectancy. A person who inherits the sickle cell gene from one parent, and a normal hemoglobin gene (HgbA) from the other, has a normal life expectancy. However, these heterozygote individuals, known as carriers of the sickle cell trait, may suffer problems from time to time.

The heterozygote is resistant to the malarial parasite which kills a large number of people each year. This is balancing selection between fierce selection against homozygous sickle-cell sufferers, and selection against the standard HgbA homozygotes by malaria. The heterozygote has a permanent advantage (a higher fitness) wherever malaria exists.

Frequency-dependent selection: Frequency-dependent selection occurs when the fitness of a phenotype is dependent on its frequency relative to other phenotypes in a given population. In positive frequency-dependent selection the fitness of a phenotype increases as it becomes more common.

In negative frequency-dependent selection the fitness of a phenotype increases as it becomes less common. For example in prey switching, rare morphs of prey are actually fitter due to predators concentrating on the more frequent morphs.

Fitness Varies in Time and Space: The fitness of a genotype may vary greatly between larval and adult stages, or between parts of a habitat range.

Selection Acts at Different Levels: The fitness of a genotype may depend on the fitness of other genotypes in the population: this covers many natural situations where the best thing to do (from the point of view of survival and reproduction) depends on what other members of the population are doing at the time.

More Complex Examples

Species in their natural habitat are often far more complex than the typical text-book examples.

Grove Snail: The Grove Snail, Cepaea nemoralis, is famous for the rich polymorphism of its shell. The system is controlled by a series of multiple alleles. Unbanded is the top dominant trait, and the forms of banding are controlled by modifier genes (see epistasis). In England the snail is regularly preyed upon by the Song Thrush Turdus philomelos, which breaks them open on thrush anvils (large stones). Here fragments accumulate, permitting researchers to analyse the snails taken. The thrushes hunt by sight, and capture selectively those forms which match the habitat least well. Snail colonies are found in woodland, hedgerows and grassland, and the predation determines the proportion of phenotypes (morphs) found in each colony.

Figure: *Grove snail, dark yellow shell with single band.*

A second kind of selection also operates on the snail, whereby certain heterozygotes have a physiological advantage over the homozygotes. Thirdly, apostatic selection is likely, with the birds preferentially taking the most common morph. This is the 'search pattern' effect, where a predominantly visual predator persists in targeting the morph which gave a good result, even though other morphs are available.

The polymorphism survives in almost all habitats, though the proportions of morphs varies considerably. The alleles controlling the polymorphism form a super-gene with linkage so close as to be nearly absolute. This control saves the population from a high proportion of undesirable recombinants.

To sum up, in this species predation by birds appears to be the main (but not the only) selective force driving the polymorphism. The snails live on heterogeneous backgrounds, and thrush are adept at detecting poor matches. The inheritance of physiological and cryptic diversity is preserved also by heterozygous advantage in the super-gene. Recent work has included the effect of shell colour on thermoregulation, and a wider selection of possible genetic influences is considered by Cook.

Chromosome polymorphism in Drosophila: In the 1930s Theodosius Dobzhansky and his co-workers collected Drosophila pseudoobscura and D. persimilis from wild populations in California and neighbouring states. Using Painter's technique they studied the polytene chromosomes and discovered that all the wild populations were polymorphic for chromosomal inversions. All the flies look alike whatever inversions they carry, so this is an example of a cryptic polymorphism. Evidence accumulated to show that natural selection was responsible:

1. Values for heterozygote inversions of the third chromosome were often much higher than they should be under the null assumption: if no advantage for any form the number of heterozygotes should conform to Ns (number in sample) = p2+2pq+q2 where 2pq is the number of heterozygotes (see Hardy-Weinberg equilibrium).
2. Using a method invented by L'Heretier and Teissier, Dobzhansky bred populations in population cages, which enabled feeding, breeding and sampling whilst preventing escape. This had the benefit of eliminating migration as a possible explanation of the results. Stocks containing inversions at a known initial frequency can be maintained in controlled conditions. It was found that the various chromosome types do not fluctuate at random, as they would if selectively neutral, but adjust to certain frequencies at which they become stabilised.
3. Different proportions of chromosome morphs were found in different areas. There is, for example, a polymorph-ratio cline in D. robusta along an 18-mile (29 km) transect near Gatlinburg, TN passing from 1,000 feet (300 m) to 4,000 feet. Also, the same areas sampled at different times of year yielded significant differences in the proportions of forms. This indicates a regular cycle of changes which adjust the population to the seasonal conditions. For these results selection is by far the most likely explanation.
4. Lastly, morphs cannot be maintained at the high levels found simply by mutation, nor is drift a possible explanation when population numbers are high.

By 1951 Dobzhansky was persuaded that the chromosome morphs were being maintained in the population by the selective advantage of the heterozygotes, as with most polymorphisms.

Heterozygote Advantage

A heterozygote advantage (heterozygous advantage) describes the case in which the heterozygote genotype has a higher relative fitness than either the homozygote dominant or homozygote recessive genotype. The specific case of heterozygote advantage is due to a single locus known as overdominance.

Polymorphism can be maintained by selection favoring the heterozygote, and this mechanism is used to explain the occurrence of some kinds of genetic variability. A common example is the case where the heterozygote conveys both advantages and disadvantages while both homozygotes convey a disadvantage. A well-established case of heterozygote advantage is that of the gene involved in sickle cell anaemia.

Often, the advantages and disadvantages conveyed are rather complicated, because more than one gene may influence a given trait or morph. Major genes almost always have multiple effects (pleiotropism), which can simultaneously convey separate advantageous traits and disadvantageous traits upon the same organism. In this instance, the state of the organism's environment will provide selection, with a net effect either favoring or working in opposition to the gene, until an environmentally-determined equilibrium is reached.

Heterozygote advantage is a major underlying mechanism for heterosis, or "hybrid vigor", which is the improved or increased function of any biological quality in a hybrid offspring.

Heterozygote Advantage in Theory

When two populations of any sexual organism are separated and kept isolated from each other, the frequencies of deleterious mutations in the two populations will differ over time, by genetic drift. It is highly unlikely, however, that the same deleterious mutations will be prevalent in both populations after a long period of separation. Since loss-of-function mutations tend to be recessive (given that dominant mutations of this type generally prevent the organism from reproducing and thereby passing the gene on to the next generation), the result of any cross between the two populations will be fitter than the parent.

This article deals with the specific case of fitness overdominance, where the fitness advantage of the cross is caused by being heterozygous at *one specific* locus alone.

Experimental Confirmation

Cases of heterozygote advantage have been demonstrated in several organisms, including humans. The first experimental confirmation of heterozygote advantage was with *Drosophila melanogaster*, a fruit fly that has been a model organism for genetic research. In a classic study, Kalmus demonstrated how polymorphism can persist in a population through heterozygote advantage.

Kalmus discovered a mutant allele of an autosomal gene that expressed ebony body colour and other selective advantages in a pattern that was autosomal dominant. The same allele also conveyed harsh disadvantages in a pattern that was completely recessive. When a fly inherited two copies of the mutation (homozygous), it expressed the dark ebony colour, but it was also particularly weak, and was placed at a harsh reproductive disadvantage.

If weakness were the only effect of the mutant allele, so that it conveyed only disadvantages, natural selection would weed out this version of the gene until it became extinct from the population. However, the same mutation also conveyed advantages, providing improved viability for individuals that were heterozygotes. The heterozygote expressed none of the disadvantages of homozygotes, yet gained improved viability. The homozygote wild type was perfectly healthy, but did not possess the improved viability of the heterozygote, and was thus at a disadvantage compared to the heterozygote in survival and reproduction.

This mutation, which at first glance appeared to be harmful, conferred enough of an advantage to heterozygotes to make it beneficial, so that it remained at dynamic equilibrium in the gene pool. Kalmus introduced flies with the ebony mutation to a wild-type population. The ebony allele persisted through many generations of flies in the study, at genotype frequencies that varied from 8% to 30%. In experimental populations, the ebony allele was more prevalent and therefore advantageous when flies were raised at low, dry temperatures, but less so in warm, moist environments.

Heterozygote Advantage in Human Genetics

Sickle-cell Anemia: Sickle-cell anemia (SCA) is a genetic disorder that is caused by the presence of two incompletely recessive alleles. When a sufferer's red blood cells are exposed to low-oxygen conditions, the cells lose their healthy round shape and become sickle-shaped. This deformation of the cells can cause them to become lodged in capillaries, depriving other parts of the body of precious oxygen. When untreated, a person with SCA may suffer from painful periodic bouts, often causing damage to internal organs, strokes, or anemia. Typically the disease results in premature death.

Because the genetic disorder is incompletely recessive, a person with only one SCA allele and one unaffected allele will have a "mixed" phenotype: The sufferer will not experience the ill effects of the disease, yet will still possess a *sickle cell trait*, whereby some of the red blood cells undergo benign effects of SCA, but nothing severe enough to be harmful. Those afflicted with sickle-cell trait are also known as carriers: If two carriers have a child, there is a twenty-five percent chance that their child will have SCA, a fifty percent chance that their child will be a carrier, and a twenty-five percent chance that the child will neither have SCA nor be a carrier. Were the presence of the SCA allele to confer only negative traits, we would expect its allele frequency to decrease generation after generation, until its presence were completely eliminated by selection and by chance.

However, there is convincing evidence indicating that, in areas with persistent malaria outbreaks, individuals with the heterozygous state have a distinct advantage (and this is why individuals with heterozygous alleles are far more common in these areas). Those with the benign sickle trait possess a resistance to malarial infection. The pathogen that causes the disease spends part of its cycle in the red blood cells, and those with sickle cells effectively stop the pathogen in its tracks, until the immune system destroys the foreign bodies. These individuals have a great immunity to infection and have a greater chance of surviving outbreaks. However, those with two alleles for SCA may survive malaria but will typically die from their genetic disease unless they have access to advanced medical care. Those of the homozygous *normal* or wild-type case will have a greater chance of passing on their genes successfully, in that there is no chance of their offspring's suffering from SCA; yet, they are more susceptible to dying from malarial infection before they have a chance to pass on their genes.

This resistance to infection is the main reason that we still see the SCA allele and SCA disease. It is found in greatest frequency in populations where malaria was and often still is a serious problem. Approximately one in 375 African-Americans is a carrier, as their recent ancestry is from malaria-stricken regions, far fewer than in Central Africa. Other populations in Africa, India, the Mediterranean and the Middle East have higher allele frequencies as well. As effective anti-malarial treatment becomes increasingly available to malaria-stricken populations, we can expect the allele frequency for SCA to decrease, so long as SCA treatments are unavailable or only partially effective. If effective Sickle-cell anemia treatments become available to the same degree, we can expect allele frequencies to remain at their present levels in these populations. In this context, 'treatment effectiveness' refers to the reproductive fitness that it grants, rather than the degree of suffering alleviation.

Cystic Fibrosis

Cystic fibrosis, or CF, is an autosomal recessive hereditary disease of the lungs, sweat glands and digestive system. The disorder is caused by the malfunction of the CFTR protein, which controls inter-membrane transport of chloride ions, which is vital to maintaining equilibrium of water in the body. The malfunctioning protein causes viscous mucus to form in the lungs and intestinal tract. Before modern times, children born with CF would have a life expectancy of only a few years, but modern medicine has made it possible for these people to live into adulthood. However, even in these individuals, male and female, CF typically causes sterility. It is the most common genetic disease among people of European descent.

The presence of a single CF mutation may influence survivorship of people affected by diseases involving loss of body fluid, typically due to diarrhea. The most common of these maladies is cholera, which throughout history has killed many Europeans. Those with cholera would often die of dehydration due to intestinal water losses. A mouse model of CF was used to study resistance to cholera, and the results were published in Science in 1994 (Gabriel, et al.). The heterozygote (carrier) mouse had less secretory diarrhea than normal, non-carrier mice. Thus it appeared for a time that resistance to cholera explained the selective advantage to being a carrier for CF and why the carrier state was so frequent.

This theory has been called into question. Hogenauer, et al. have challenged this popular theory with a human study. Prior data were based on solely on mouse experiments. These authors found that the heterozygote state was indistinguishable from the non-carrier state.

Another theory for the prevalence of the CF mutation is that it provides resistance to tuberculosis. Tuberculosis was responsible for 20% of all European deaths between 1600 and 1900, so even partial protection against the disease could account for the current gene frequency.

As of 2007, the selective pressure for the high gene prevalence of CF mutations is still uncertain. Approximately 1 in 25 persons of European descent is a carrier of the disease, and 1 in 2500 to 3000 children born is affected by cystic fibrosis.

Triosephosphate Isomerase

Triosephosphate isomerase (TPI) is a central enzyme of glycolysis, the main pathway for cells to obtain energy metabolizing carbon sugars. In humans, certain mutations within this enzyme which affect the dimerisation of this protein are causal for a rare disease, triosephosphate isomerase deficiency. Other mutations, which inactivate the enzyme (= null alleles) are lethal when inherited homozygously (two defect copies of the TPI gene), but have no obvious effect as heterozygotes (one defect and one normal copy). However, the frequency of heterozygous null-alleles is much higher than expected, indicating a heterozygous advantage for TPI null alleles as well. The reason is unknown, however, new scientific results are suggesting that cells having reduced TPI activity are more resistant against oxidative stress PlosOne, Dec. 2006

Resistance to Hepatitis C Virus Infection

There is evidence that genetic heterozygosity in humans provides increased resistance to certain viral infections. There is significantly lower proportions of HLA-DRB1 heterozygosity among HCV-infected than uninfected cases. The differences were more pronounced with alleles represented as functional supertypes (P = 1.05 x 10(-6)) than as low-resolution genotypes (P = 1.99 x 10(-3)). These findings constitute evidence that heterozygosity provides an advantage among carriers of different supertype HLA-DRB1 alleles against HCV infection progression to end-stage liver disease in a large-scale, long-term study population.

MHC Heterozygosity and Human scent Preferences

Multiple studies have shown that, in double-blind experiments, women prefer the scent of men who are heterozygous at all three MHC loci. The reasons proposed for these findings are speculative, however, it has been argued that heterozygosity at MHC loci results in more alleles to fight against a wider variety of diseases, possibly increasing survival rates against a wider range of infectious diseases. The latter claim has been tested in an experiment, which showed that outbreeding mice to exhibit MHC heterozygosity enhanced their health and survival rates against multiple-strain infections.

Ashkenazi Jewish Intelligence

A more recent theory of heterozygote advantage detailed in the 2005 paper "Natural History of Ashkenazi Intelligence" proposes that Tay-Sachs, and the other lipid storage diseases that are prevalent in Ashkenazi Jews, reflect genes that enhance dendrite growth and promote higher intelligence in Ashkenazi Jews when present in carrier form.

Frequency-dependent Selection

Frequency-dependent selection is the term given to an evolutionary process where the fitness of a phenotype is dependent on its frequency relative to other phenotypes in a given population. In positive frequency-dependent selection (or purifying frequency-dependent selection) the fitness of a phenotype increases as it becomes more common. In negative frequency-dependent selection (or diversifying frequency-dependent selection) the fitness of a phenotype increases as it becomes rarer. Negative frequency-dependent selection is an example of balancing selection. Frequency-dependent selection is usually the result of interactions between species (predation, parasitism, or competition) or between genotypes within species (usually competitive or symbiotic), and has been especially frequently discussed with relation to anti-predator adaptations. Frequency-dependent selection can lead to polymorphic equilibria which result from interactions among genotypes within species in the same way that multi-species equilibria require interactions between species in competition (e.g. where $\acute{a}_{ij}$ parameters in Lotka-Volterra competition equations are non-zero).

Negative Frequency-dependent Selection

The first explicit statement of frequency dependent selection appears to have been by E.B. Poulton in 1884 with reference to the way that predators could maintain colour polymorphisms in their prey. Perhaps the best known early modern statement of the principle in the twentieth century was the discussion by Bryan Clarke of apostatic selection (a synonym of negative frequency-dependent selection). In this paper, Clarke discussed frequency-dependent selection particularly with regard to predator attacks on polymorphic British snails, citing Luuk Tinbergen's classic work on searching images as support that predators such as birds would tend to specialise on common forms of palatable species. Clarke later became a major promoter of the idea that frequency-dependent balancing selection might explain abundant molecular polymorphisms (often in the absence of heterosis) in opposition to the neutral theory of molecular evolution.

Another example of negative frequency-dependent selection is in the case of plant self-incompatibility alleles. When two plants share the same incompatibility allele, they are unable to mate. Thus, a plant with a new (and therefore, rare) allele has more success at mating, and its allele spreads quickly through the population.

Negative frequency-dependent selection also operates in the interaction of many human pathogens, such as the flu virus. Once a particular strain has been common in a human population, most individuals would have developed an immune response to that strain. But a rare, novel strain of the flu virus would be able to spread quickly to almost any individual. This advantage of genetic novelty causes continual evolution of viral strains, with new versions common each year. Another immune-related example of negative frequency-dependent selection is the major histocompatibility complex, which is involved in recognition of foreign antigens). Frequency-dependent selection as an important factor may explain the high degree of polymorphism of MHC (Borghans et al., 2004).

Positive Frequency-dependent Selection

Where negative frequency-dependent selection gives an advantage to rare phenotypes, positive frequency-dependent selection gives an advantage to common phenotypes. In the between-species analogue, this is equivalent to an Allee effect, in which if a species is too rare, it may decline to extinction. This means that new alleles can have a difficult time invading a population, since they don't experience significant benefit until they become common. This has been proposed as a difficulty in the evolution of aposematic (or warning) colouration in toxic or distasteful organisms. The presumed advantage of the aposematic colouration is that predators have learned to avoid potential prey with that colour pattern. But when the pattern is rare, the predator population does not become 'educated' and the pattern has no benefit. Therefore the warning colour is only advantageous once it has become common. Warning colouration, if it involves more than one species, is known as Müllerian mimicry, a form of convergent evolution where multiple species share the same advantageous pattern.

More Complex Examples

Species in their natural habitat are often far more complex than the typical text-book examples.

Grove Snail

The Grove Snail, *Cepaea nemoralis,* is famous for the rich polymorphism of its shell. The system is controlled by a series of multiple alleles. Unbanded is the top dominant trait, and the forms of banding are controlled by modifier genes (see epistasis).

In England the snail is regularly preyed upon by the Song Thrush *Turdus philomelos*, which breaks them open on *thrush anvils* (large stones). Here fragments accumulate, permitting researchers to analyse the snails taken. The thrushes hunt by sight, and capture selectively those forms which match the habitat *least well.* Snail colonies are found in woodland, hedgerows and grassland, and the predation determines the proportion of phenotypes (morphs) found in each colony.

A second kind of selection also operates on the snail, whereby certain heterozygotes have a physiological advantage over the homozygotes. Thirdly, apostatic selection is likely, with the birds preferentially taking the most common morph. This is the 'search pattern' effect, where a predominantly visual predator persists in targeting the morph which gave a good result, even though other morphs are available.

The polymorphism survives in almost all habitats, though the proportions of morphs varies considerably. The alleles controlling the polymorphism form a super-gene with linkage so close as to be nearly absolute. This control saves the population from a high proportion of undesirable recombinants.

To sum up, in this species predation by birds appears to be the main (but not the only) selective force driving the polymorphism. The snails live on heterogeneous backgrounds, and thrush are adept at detecting poor matches. The inheritance of physiological and cryptic diversity is preserved also by heterozygous advantage in the super-gene. Recent work has included the effect of shell colour on thermoregulation, and a wider selection of possible genetic influences is considered by Cook.

Chromosome Polymorphism in Drosophila

In the 1930s Theodosius Dobzhansky and his co-workers collected *Drosophila pseudoobscura* and *D. persimilis* from wild populations in California and neighbouring states. Using Painter's technique they studied the polytene chromosomes and discovered that all the wild populations were polymorphic for chromosomal inversions. All the flies look alike whatever inversions they carry, so this is an example of a cryptic polymorphism. Evidence accumulated to show that natural selection was responsible:

1. Values for heterozygote inversions of the third chromosome were often much higher than they should be under the null assumption: if no advantage for any form the number of heterozygotes should conform to N_s (number in sample) $= p+2pq+q$ where 2pq is the number of heterozygotes.
2. Using a method invented by L'Heretier and Teissier, Dobzhansky bred populations in *population cages*, which enabled feeding, breeding and sampling whilst preventing escape. This had the benefit of eliminating migration as a possible explanation of the results. Stocks containing inversions at a known initial frequency can be maintained in controlled conditions. It was found that the various chromosome types do not fluctuate at random, as they would if selectively neutral, but adjust to certain frequencies at which they become stabilised.

3. Different proportions of chromosome morphs were found in different areas. There is, for example, a polymorph-ratio cline in *D. robusta* along an 18-mile (29 km) transect near Gatlinburg, TN passing from 1,000 feet (300 m) to 4,000 feet. Also, the same areas sampled at different times of year yielded significant differences in the proportions of forms. This indicates a regular cycle of changes which adjust the population to the seasonal conditions. For these results selection is by far the most likely explanation.
4. Lastly, morphs cannot be maintained at the high levels found simply by mutation, nor is drift a possible explanation when population numbers are high.

By 1951 Dobzhansky was persuaded that the chromosome morphs were being maintained in the population by the selective advantage of the heterozygotes, as with most polymorphisms.

Directional Selection

In population genetics, directional selection is a mode of natural selection in which a single phenotype is favored, causing the allele frequency to continuously shift in one direction. Under directional selection, the advantageous allele increases in frequency independently of its dominance relative to other alleles; that is, even if the advantageous allele is recessive, it will eventually become fixed. Directional selection stands in contrast to balancing selection where selection may favourmultiple alleles, and is the same as purifying selection which removes deleterious mutations from a population.

Directional selection occurs most often under environmental changes and when populations migrate to new areas with different environmental pressures. An example of directional selection is fossil records that show that the size of the black bears in Europe decreased during interglacial periods of the ice ages, but increased during each glacial period.

Another example is the beak size in a population of finches. Throughout the wet seasons, small seeds were more common and there was such a large supply of the small seeds that the finches rarely ate large seeds. During the dry seasons, none of the seeds were in great abundance, but the birds usually ate more large seeds. The change in diet of the finches affected the depth of the birds' beaks in the future generations.Their beaks range from large and tough to small and smooth.

Disruptive Selection

Disruptive selection, also called diversifying selection, describes changes in population genetics in which extreme values for a trait are favored over intermediate values. In this case, the variance of the trait increases and the population is divided into two distinct groups. This evolutionary process is believed to be the driving force behind sympatric speciation.

Example

Suppose there is a population of rabbits. The colour of the rabbits is governed by two incompletely dominant traits: black fur represented by "B" and white fur represented by "b". A rabbit with the genotype of "BB" would have a phenotype of black fur, a genotype of "Bb" would have grey fur (a display of both black and white) and a genotype of "bb" would have a phenotype of white fur.

If this population of rabbits were put into an area that had very dark black rocks as well as very white coloured stone, the rabbits with black fur would be able to hide from predators amongst the black rocks and the white furred rabbits would be able to hide in the white rocks, but the grey furred rabbits would stand out in both of the habitats and would suffer greater predation.

As a consequence of the selective pressures of their environment, our hypothetical rabbit population would be disruptively selected for extreme values of the fur colour trait: white or black, but not grey.

Significance

Disruptive selection is of particular significance in the history of evolutionary study, as it is involved in one of evolution's "cardinal cases", namely the finch populations observed by Darwin in the Galápagos.

He observed that the species of finches were similar enough to ostensibly have been descended from a single species. However, they exhibited disruptive variation in beak size. This variation appeared to be adaptively related to the seed size available on the respective islands (big beaks for big seeds, small beaks for small seeds). Medium beaks had difficulty retrieving small seeds and were also not tough enough for the bigger seeds, and were hence maladaptive.

While it is true that disruptive selection can lead to speciation, this is not as quick or straightforward of a process as other types of speciation or evolutionary change. This is largely because the results of disruptive selection are less stable than the results of directional selection (directional selection favours individuals at only one end of the spectrum).For example, let us take the mathematically straightforward yet biologically improbable case of the rabbits: Suppose directional selection were taking place. The field only has dark rocks in it, so the darker the rabbit, the better. Eventually there will be a lot of black rabbits in the population (hence lots of "B" alleles) and a lesser amount of gray rabbits (who contribute 50% chromosomes with "B" allele and 50% chromosomes with "b" allele to the population). There will be few white rabbits (not very many contributors of chromosomes with "b" allele to the population). This could eventually lead to a situation in which chromosomes with "b" allele die out, making black the only possible colour for all subsequent rabbits. The reason for this is that there is nothing "boosting" the level of "b" chromosomes in the population. They can only go down, and eventually die out.

Consider now the case of disruptive selection. The result is equal numbers of black and white rabbits, and hence equal numbers of chromosomes with "B" or "b" allele, still floating around in that population. Every time a white rabbit mates with a black one, only gray rabbits results. So, in order for the results to "click", there needs to be a force causing white rabbits to choose other white rabbits, and black rabbits to choose other black ones. In the case of the finches, this "force" was geographic/niche isolation.

Stabilising Selection

Description

Stabilising or ambidirectional selection, (not the same thing as negative or purifying selection), is a type of natural selection in which genetic diversity decreases as the population stabilises on a particular trait value. This is probably the most common mechanism of action for natural selection. Stabilising selection commonly uses negative selection (a.k.a. purifying selection) to select against extreme values of the character. Stabilising Selection is the opposite of disruptive selection, instead of favoring individuals with extreme phenotypes, it favors the intermediate variants. It reduces phenotypic variation and maintains the status quo. Natural selection tends to remove the more severe phenotypes, resulting in the reproductive success of the norm or average phenotypes.

A classic example of this is human birth weight. Babies of low weight lose heat more quickly and get ill from infectious disease more easily, whereas babies of large body weight are more difficult to deliver through the pelvis. Infants of a more medium weight survive much more often. For the larger or smaller babies, the baby mortality rate is much higher. Stabilising selection operates most of the time in most populations. This type of selection acts to prevent divergence of form and function. In this way, the anatomy of some organisms, such as sharks and ferns, has remained largely unchanged for millions of years.

Stabilising selection can sometimes be detected by measuring the fitness of the range of different phenotypes by various direct measures, but it can also be detected by a variety of tests of molecular sequence data, such as Ka/Ks ratios, changes in allele frequency distributions, and the McDonald-Kreitman test.

In human populations, 5% of the genome DNA is under purified selection for common mammalian genes. More loci may rapidly evolve including regulatory sequences and miRNAs.

Parental Investment

In evolutionary biology, parental investment (PI) is any parental expenditure (time, energy etc.) that benefits one offspring at a cost to parents' ability to invest in other components of fitness (Clutton-Brock 1991: 9; Trivers 1972). Components of fitness (Beatty 1992) include the wellbeing of existing offspring, parents' future reproduction, and inclusive fitness through aid to kin (Hamilton, 1964). Parental investment is sometimes incorrectly equated with parental care or parental effort. What is the difference?

Parental investment theory is a branch of life history theory. The earliest consideration of parental investment is given by Fisher (1930), Fisher's principle, wherein Fisher argued that parental expenditure on both sexes of offspring should be equal. Clutton-Brock (1991: 9) expanded the concept of PI to include costs to any other component of parental fitness. Robert Trivers' theory of parental investment predicts that the sex making the largest investment in lactation, nurturing, and protecting offspring will be more discriminating in mating; and that the sex that invests less in offspring will compete for access to the higher-investing sex (see Bateman's principle). Sex differences in parental effort are important in determining the strength of sexual selection.

Reproduction is costly. Individuals are limited in the degree to which they can devote time and resources to producing and raising their young, and such expenditure may also be detrimental to their future condition, survival, and further reproductive output. However, such expenditure is typically beneficial to the offspring, enhancing their condition, survival, and reproductive success. These differences may lead to parent-offspring conflict. Parental investment can be provided by the female (female uniparental care), the male (male uniparental care), or both (biparental care). Parents are naturally selected to maximise the difference between the benefits and the costs, and parental care will tend to exist when the benefits are substantially greater than the costs.

Parental care is found in a broad range of taxonomic groups, including both ectothermic (invertebrates, fish, amphibians and reptiles), and endothermic (birds and mammals) species. Care can be provided at any stage of the offspring's life: pre-natal care including behaviours such as egg guarding, preparation of nest, brood carrying, incubation, and placental nourishment in mammals; and post-natal care including food provisioning and protection of offspring.

In iteroparous species, where individuals may go through several reproductive bouts during their lifetime, a tradeoff may exist between investment in current offspring and future reproduction. Parents need to balance their offspring's demands against their own self-maintenance. This potential negative effect of parental care was explicitly formalised by Trivers (1972), who originally defined the term *parental investment* to mean *any investment by the parent in an individual offspring that increases the offspring's chance of surviving (and hence reproductive success) at the cost of the parent's ability to invest in other offspring.*

The benefits of parental investment to the offspring are large and are associated with the effects on condition, growth, survival, and ultimately on reproductive success of the offspring. However, these benefits can come at the cost of parent's ability to reproduce in the future e.g., through increased risk of injury when defending offspring against predators, loss of mating opportunities whilst rearing offspring, and an increase in the time interval until the next reproduction.

Overall, parents are selected to maximise the difference between the benefits and the costs, and parental care will be likely to evolve when the benefits exceed the costs.

Ethology

Ethology is the scientific study of animal behaviour, and a sub-topic of zoology.

Although many naturalists have studied aspects of animal behaviour throughout history, the modern discipline of ethology is generally considered to have begun during the 1930s with the work of Dutch biologist Nikolaas Tinbergen and Austrian biologists Konrad

Lorenz and Karl von Frisch, joint winners of the 1973 Nobel Prize in Physiology or Medicine.

Ethology is a combination of laboratory and field science, with a strong relation to certain other disciplines such as neuroanatomy, ecology, and evolution. Ethologists are typically interested in a behavioural process rather than in a particular animal group, and often study one type of behaviour (e.g. aggression) in a number of unrelated animals.

The desire to understand animals has made ethology a rapidly growing field. Since the turn of the 21st century, many aspects of animal communication, personal symbolic name use, animal emotions, animal culture, learning, and even sexual conduct that experts long thought they understood, have been reexamined, and new conclusions reached. New fields have developed, such as neuroethology.

Etymology

The term *ethology* derives from the Greek word meaning *character*. Other words that derive from *ethos" include* ethics *and* ethical. *The term was first popularised by American myrmecologist William Morton Wheeler in 1902. An earlier, slightly different sense of the term was proposed by John Stuart Mill in his 1843 System of Logic. He recommended the development of a new science, "ethology," the purpose of which would be explanation of individual and national differences in character, on the basis of associationistic psychology. This use of the word was never adopted.*

Relationship with Comparative Psychology

Comparative psychology also studies animal behaviour, but, as opposed to ethology, is construed as a sub-topic of psychology rather than as one of biology. Historically, where comparative psychology researches animal behaviour in the context of what is known about human psychology, ethology researches animal behaviour in the context of what is known about animal anatomy, physiology, neurobiology, and phylogenetic history. Furthermore, early comparative psychologists concentrated on the study of learning and tended to research behaviour in artificial situations, whereas early ethologists concentrated on behaviour in natural situations, tending to describe it as instinctive. The two approaches are complementary rather than competitive, but they do result in different perspectives and, sometimes, conflicts of opinion about matters of substance. In addition, for most of the twentieth century, comparative psychology developed most strongly in North America, while ethology was stronger in Europe. A practical difference is that early comparative psychologists concentrated on gaining extensive knowledge of the behaviour of very few species. Ethologists were more interested in understanding behaviour in a wide range of species to facilitate principled comparisons across taxonomic groups. Ethologists have made much more use of a truly comparative method than comparative psychologists have. Despite the historical divergence, most ethologists, at least in North America, teach in psychology departments. It is a strong belief among scientists that the mechanisms on which behavioural processes are based are the same that cause the evolution of the living species: there is therefore a strong association between these two fields.

Scala Naturae and Lamarck's Theories

Until the 19th century, the most common theory among scientists was still the concept of *scala naturae*, proposed by Aristotle. According to this theory, living beings were classified on an ideal pyramid that represented the simplest animals on the lower levels, with complexity increasing progressively toward the top, occupied by human beings. In the Western world of the time, people believed animal species were eternal and immutable, created with a specific purpose, as this seemed the only possible explanation for the incredible variety of living beings and their surprising adaptation to their habitats.

Jean-Baptiste Lamarck (1744 - 1829) was the first biologist to describe a complex theory of evolution. His theory substantially comprised two statements: first, that animal organs and behaviour can change according to the way they are used; and second, that those characteristics can transmit from one generation to the next (the example of the giraffe whose neck becomes longer while trying to reach the upper leaves of a tree is well-known). The second statement is that every living organism, humans included, tends to reach a greater level of perfection. When Charles Darwin went to the Galapagos Islands, he was well aware of Lamarck's theories and was influenced by them.

Theory of Evolution by Natural Selection and the Beginnings of Ethology

Because ethology is considered a topic of biology, ethologists have been concerned particularly with the evolution of behaviour and the understanding of behaviour in terms of the theory of natural selection. In one sense, the first modern ethologist was Charles Darwin, whose book, *The Expression of the Emotions in Man and Animals*, influenced many ethologists. He pursued his interest in behaviour by encouraging his protégé George Romanes, who investigated animal learning and intelligence using an anthropomorphic method, anecdotal cognitivism, that did not gain scientific support.

Other early ethologists, such as Oskar Heinroth and Julian Huxley, instead concentrated on behaviours that can be called instinctive, or natural, in that they occur in all members of a species under specified circumstances. Their beginning for studying the behaviour of a new species was to construct an ethogram (a description of the main types of natural behaviour with their frequencies of occurrence). This provided an objective, cumulative base of data about behaviour, which subsequent researchers could check and supplement.

Fixed Action Patterns and Animal Communication

An important development, associated with the name of Konrad Lorenz though probably due more to his teacher, Oskar Heinroth, was the identification of fixed action patterns (FAPs). Lorenz popularised FAPs as instinctive responses that would occur reliably in the presence of identifiable stimuli (called sign stimuli or releasing stimuli). These FAPs could then be compared across species, and the similarities and differences between behaviour could be easily compared with the similarities and differences in morphology.

An important and much quoted study of the Anatidae (ducks and geese) by Heinroth used this technique. Ethologists noted that the stimuli that released FAPs were commonly features of the appearance or behaviour of other members of the animal's own species, and they were able to prove how important forms of animal communication could be mediated by a few simple FAPs. The most sophisticated investigation of this kind was the study by Karl von Frisch of the so-called "dance language" related to bee communication. Lorenz developed an interesting theory of the evolution of animal communication based on his observations of the nature of fixed action patterns and the circumstances in which animals emit them.

Instinct

The Merriam-Webster dictionary defines instinct as a largely inheritable and unalterable tendency of an organism to make a complex and specific response to environmental stimuli without involving reason. For ethologists, instinct means a series of predictable behaviours for fixed action patterns. Such schemes are only acted when a precise stimulating signal is present. When such signals act as communication among members of the same species, they are known as releasers. A notable example of a releaser is the beak movements in many bird species performed by the newborns, which stimulates the mother's regurgitating process to feed her offspring. Another well-known case is the classic experiments by Tinbergen on the Graylag Goose.

Like similar waterfowl, the goose rolls a displaced egg near its nest back to the others with its beak. The sight of the displaced egg triggers this mechanism. If the egg is taken away, the animal continues with the behaviour, pulling its head back as if an imaginary egg is still being maneuvered by the underside of its beak. However, it also attempts to move other egg-shaped objects, such as a giant plaster egg, door knob, or even a volleyball back into the nest. Such objects, when they exaggerate the releasers found in natural objects, can elicit a stronger version of the behaviour than the natural object, so that the goose ignores its own displaced egg in favourof the giant dummy egg. These exaggerated releasers for instincts were named supernormal stimuli by Tinbergen. Tinbergen found he could produce supernormal stimuli for most instincts in animals—such as cardboard butterflies that male butterflies preferred to mate with if they had darker stripes than a real female, or dummy fish that a territorial male stickleback fish fought more violently than a real invading male if the dummy had a brighter-coloured underside. Harvard psychologist Deirdre Barrett has done research pointing out how easily humans also respond to supernormal stimuli for sexual, nurturing, feeding, and social instincts. However, a behaviour only made of fixed action patterns would be particularly rigid and inefficient, reducing the probability of survival and reproduction, so the learning process has great importance, as does the ability to change the individual's responses based on its experience. It can be said that the more the brain is complex and the life of the individual long, the more its behaviour is "intelligent" (in the sense of being guided by experience rather than stereotyped FAPs).

Learning

Learning occurs in many ways, one of the most elementary being habituation. This process consists of ignoring persistent or useless stimuli. An example of learning by habituation is the one observed in squirrels: When one of them feels threatened, the others hear its signal and go to the nearest refuge. However, if the signal comes from an individual that has caused many false alarms, the other squirrels ignore the signal. Another common way of learning is by association, where a stimulus is, based on the experience, linked to another one that may not have anything to do with the first one. The first studies of associative learning were made by Russian physiologist Ivan Pavlov. An example of associative behaviour is observed when a common goldfish goes close to the water surface whenever a human is going to feed it, or the excitement of a dog whenever it sees a collar as a prelude for a walk.

Imprinting

Imprinting is the term used in psychology and ethology to describe any kind of phase-sensitive learning (learning occurring at a particular age or a particular life stage) that is rapid and apparently independent of the consequences of behaviour. It was first used to describe situations in which an animal or person learns the characteristics of some stimulus, which is therefore said to be "imprinted" onto the subject. Imprinting is hypothesized to have a critical period.

Filial Imprinting

The best-known form of imprinting is *filial imprinting,* in which a young animal acquires several of its behavioural characteristics from its parent. It is most obvious in nidifugous birds, which imprint on their parents and then follow them around. It was first reported in domestic chickens, by the 19th-century amateur biologist Douglas Spalding. It was rediscovered by the early ethologist Oskar Heinroth, and studied extensively and popularised by his disciple Konrad Lorenz working with greylag geese. Lorenz demonstrated how incubator-hatched geese would imprint on the first suitable moving stimulus they saw within what he called a "critical period" between 13–16 hours shortly after hatching. For example, the goslings would imprint on Lorenz himself (to be more specific, on his wading boots), and he is often depicted being followed by a gaggle of geese who had imprinted on him. Filial imprinting is not restricted to non-human animals that are able to follow their parents, however. In child development, the term is used to refer to the process by which a baby learns who its mother and father are. The process is recognised as beginning in the womb, when the unborn baby starts to recognise its parents' voices.

The filial imprinting of birds was a primary technique used to create the movie *Winged Migration (Le Peuple Migrateur),* which contains a great deal of footage of migratory birds in flight. The birds imprinted on handlers, who wore yellow jackets and honked horns constantly. The birds were then trained to fly along with a variety of aircraft, primarily ultralights.

The Italian hang-glider pilot Angelo d'Arrigo extended this technique. D'Arrigo noted that the flight of a non-motorised hang-glider is very similar to the flight patterns of migratory birds: Both use updrafts of hot air (thermal currents) to gain altitude that then permits soaring flight over distance. He used this fact to enable the reintroduction into the wild of threatened species of raptors. Birds that are hatched in captivity have no mentor birds to teach them their traditional migratory routes. D'Arrigo had one solution to this problem. The chicks hatched under the wing of his glider, and imprinted on him. Then, he taught the fledglings to fly and to hunt. The young birds followed him not only on the ground (as with Lorenz) but also in the air as he took the path of various migratory routes. He flew across the Sahara and over the Mediterranean Sea to Sicily with eagles, from Siberia to Iran (5,500 km) with a flock of Siberian cranes, and over Mount Everest with Nepalese eagles. In 2006, he worked with a condor in South America.

In a similar project, orphaned Canada Geese were trained to their normal migration route by the Canadian ultralight enthusiast Bill Lishman, as shown in the fact-based movie drama *Fly Away Home*.

Chicks of domestic chickens prefer to be near large groups of objects that they have imprinted on. This behaviour was used to determine that very young chicks of a few days old have rudimentary counting skills. In a series of experiments, they were made to imprint on plastic balls and could figure out which of two groups of balls hidden behind screens had the most balls.

Sexual Imprinting

Sexual imprinting is the process by which a young animal learns the characteristics of a desirable mate. For example, male zebra finches appear to prefer mates with the appearance of the female bird that rears them, rather than mates of their own type.

Sexual imprinting on inanimate objects is a popular theory concerning the development of sexual fetishism. For example, according to this theory, imprinting on shoes or boots (as with Konrad Lorenz' geese) would be the cause of shoe fetishism.

Westermarck Effect

Reverse sexual imprinting is also seen in instances where two people who live in close domestic proximity during the first few years in the life of either one become desensitised to later close sexual attraction. This phenomenon, known as the Westermarck effect, was first formally described by Finnish anthropologist Edvard Westermarck in his book *The History of Human Marriage* (1891). The Westermarck effect has since been observed in many places and cultures, including in the Israeli kibbutz system, and the Chinese Shim-pua marriage customs, as well as in biological-related families.

In the case of the Israeli kibbutzim (collective farms), children were reared somewhat communally in peer groups, based on age, not biological relation. A study of the marriage patterns of these children later in life revealed that out of the nearly 3,000 marriages that occurred across the kibbutz system, only fourteen were between children from the same peer group. Of those fourteen, none had been reared together during the first six years of life. This result provides evidence not only that the Westermarck effect is demonstrable but that it operates during the period from birth to the age of six.

When proximity during this critical period does not occur—for example, where a brother and sister are brought up separately, never meeting one another—they may find one another highly sexually attractive when they meet as adults. This phenomenon is known as genetic sexual attraction. This observation supports the hypothesis that the Westermarck effect evolved because it suppressed inbreeding. This attraction may also be seen with cousin couples.

Westermarck and Freud

Freud argued that as children, members of the same family naturally lust for one another, making it necessary for societies to create incest taboos, but Westermarck argued the reverse, that the taboos themselves arise naturally as products of innate attitudes.

Steven Pinker wrote on the subject:

The idea that boys want to sleep with their mothers strikes most men as the silliest thing they have ever heard. Obviously, it did not seem so to Freud, who wrote that as a boy he once had an erotic reaction to watching his mother dressing. But Freud had a wet-nurse, and may not have experienced the early intimacy that would have tipped off his perceptual system that Mrs. Freud was his mother. The Westermarck theory has *out-Freuded* Freud.—Steven Pinker, *How the Mind Works*

Imitation

Imitation is often an important type of learning. A well-documented example of imitative learning occurred in a group of macaques on Hachijojima Island, Japan. The macaques lived in the inland forest until the 1960s, when a group of researchers started giving them potatoes on the beach: soon, they started venturing onto the beach, picking the potatoes from the sand, and cleaning and eating them. About one year later, an individual was observed bringing a potato to the sea, putting it into the water with one hand, and cleaning it with the other. Her behaviour was soon imitated by the individuals living in contact with her; when they gave birth, they taught this practice to their young. The National Institutes of Health reported that capuchin monkeys preferred the company of researchers who imitated them to that of researchers who did not. The monkeys not only spent more time with their imitators but also preferred to engage in a simple task with them even when provided with the option of performing the same task with a non-imitator.

Mating and the Fight for Supremacy

Individual reproduction is the most important phase in the proliferation of individuals or genes within a species: for this reason, there exist complex mating rituals, which can be very complex even if they are often regarded as fixed action patterns (FAPs). The Stickleback's complex mating ritual was studied by Niko Tinbergen and is regarded as a notable example of a FAP.

Often in social life, animals fight for the right to reproduce, as well as social supremacy. A common example of fighting for social and sexual supremacy is the so-called pecking order among poultry. Every time a group of poultry cohabitate for a certain time length, they establish a pecking order. In these groups, one chicken dominates the others and can peck without being pecked. A second chicken can peck all the others except the first, and so on. Higher level chickens are easily distinguished by their well-cured aspect, as opposed to lower level chickens. While the pecking order is establishing, frequent and violent fights can happen, but once established, it is broken only when other individuals enter the group, in which case the pecking order re-establishes from scratch.

Bibliography

Alger, Janet and Steven Alger: *Cat Culture: The Social World of a Cat Shelter,* Temple University Press, Philadelphia, 2003.

Allen, Karen Miller: *The Human-Animal Bond: An Annotated Bibliography,* Scarecrow Press, Metuchen, NJ, 1985.

Baldick, Julian: *Animals and Shaman: Ancient Religions of Central Asia,* New York University Press, New York, 2000.

Bauston, Gene: *Battered Birds, Crated Herds: How We Treat the Animals We Eat,* Farm Sanctuary, Watkins Glen, NY, 1996.

Calarco, Matthew, and Peter Atterton: *Animal Philosophy: Essential Readings in Continental Thought,* Continuum, London, 2004.

Clutton-Brock, Juliet: *Domesticated Animals from Early Times,* University of Texas Press, Austin, TX, 1981.

DeGrazia, David: *Taking Animals Seriously: Mental Life and Moral Status,* Cambridge, New York, 1996.

Dolins, Francine: *Attitudes to Animals: Views on Animal Welfare,* Cambridge University Press, Cambridge, 1999.

Fine, Aubrey H.: *Handbook on Animal-Assisted Therapy: Theoretical Foundations and Guidelines for Practice,* Academic Press, UK, 2006.

Fox, Michael Allen: *The Case for Animal Experimentation,* University of California Press, Berkeley, CA, 1986.

Gates, P.: *Animal Communication,* Cambridge University Press, Cambridge, 1997.

Harbolt-Bosco, Tami: *Bridging the Bond: The Cultural Construction of the Shelter Pig,* Purdue University Press, 2003.

Harris, Marvin: *The Sacred Cow and the Abominable Pig: Riddles of Food and Culture,* Touchstone Books, New York, 1987.

Jones, Susan D.: *Valuing Animals: Veterinarians and Their Patients in Modern America,* Johns Hopkins University Press, Baltimore, MD, 2003.

Katz, Jon: *The New Work of Dogs: Tending to Life, Love, and Family,* Villard, New York, 2003.

Langley , Gill: *Animal Experimentation: The Consensus Changes*, Chapman and Hall, New York, 1989.

Lilequist , J.: *Forbidden History, the State, Society and the Regulation of Sexuality in Modern Europe,* Chicago University Press, Chicago, 1992.

Masson, Jeffrey Moussaieff: *Dogs Never Lie About Love: Reflections on the Emotional World of Dogs*, Crown Publishers, New York, 1997.

Mitchell, W.J.T.: *The Last Dinosaur Book: The Life and Times of a Cultural Icon,* University of Chicago Press, Chicago and London, 1998.

Noske, Barbara: *Humans and Other Animals: Beyond the Boundaries of Anthropology,* Pluto Press, London, 1989.

Phillips, Ian: *Lost: Lost and Found Pet Posters from Around the World,* Princeton Architectural Press, New York, 2002.

Rifkin, Jeremy: *Beyond Beef: The Rise and Fall of the Cattle Culture*, Dutton, New York, 1992.

Ristau, C.: *Cognitive Ethology: The Minds of Other Animals*, Lawrence Erlbaum, Hillsdale, NJ, 1990.

Rogers, Katherine: *The Cat and the Human Imagination,* The University of Michigan Press, Ann Arbor, 2001.

Salem, Harry: *Animal Test Alternatives: Refinement, Reduction, Replacement,* M. Dekker, New York, 1995.

Sharpe, Robert: *Science on Trial: The Human Cost of Animal Experiments*, Awareness Books, Sheffield, UK, 1994.

Tabor, Roger K.: *The Wild Life of the Domestic Cat,* Arrow Books, London, 1983.

Taylor, R.E. and R. Bogart: *Scientific Farm Animal Production: An Introduction to Animal Science*, MacMillan, New York, 1988.

Thomas, Elizabeth Marshall: *The Tribe of the Tiger,* Simon and Schuster, New York, 1994.

Topsells, E.: *History of Four-footed Beast,* Nelson Hall, Chicago, 1981.

Ucko, Peter J. and G.W. Dimbleby: *The Domestication and Exploitation of Plants and Animals*, Aldine Publishing Company, Chicago, IL, 1969.

Willis, R.G.: *Signifying Animals: Human Meaning in the Natural World,* Unwin Hyman, London, 1990.

Yarri, Donna: *The Ethics of Animal Experimentation: A Critical Analysis And Constructive Christian Proposal,* Oxford University Press, Oxford, 2005.

Zeuner, Frederick E.: *A History of Domesticated Animals*, Hutchinson, London, 1963.

Index

❑❑❑